I0047940

William Paul Gerhard

Hints on the Drainage of Dwellings

William Paul Gerhard

Hints on the Drainage of Dwellings

ISBN/EAN: 9783744678360

Printed in Europe, USA, Canada, Australia, Japan

Cover: Foto ©berggeist007 / pixelio.de

More available books at **www.hansebooks.com**

HINTS

ON THE

Drainage and Sewerage of Dwellings.

BY

WM. PAUL GERHARD,

CIVIL ENGINEER.

NEW YORK:

WILLIAM T. COMSTOCK,

6 ASTOR PLACE.

1884.

PREFACE.

This little work has grown out of a series of articles, contributed by the author, under the pseudonym "Hippocrates," to the columns of "BUILDING."

It has been the author's aim to give an account of the usual condition in which plumbing work, done years ago—and some done quite recently—may be found, and also to give suggestions on the proper manner of doing the work. The title "Hints"

ERRATA.

Page 6 line 6 from the top, read "preventable" instead of "preventive."

" 17 line 9 " " " " "filth."

" 19 bottom line "use of the closet" instead of "use of closet."

" 134 line 22 from the top read "from the outlet" instead of "from the soil pipe."

" 156 line 9 from the top read "known" instead of "knows."

" 179 " 19 " " " " "free from" " " "free of."

" 194 " 12 " " bottom read "by heating and then dipping them."

" 246 " 8 " " top read "my" instead of "any."

" 248 " 2 " " " " "movable" instead of "marble."

New York, January 1st, 1884.

PREFACE.

This little work has grown out of a series of articles, contributed by the author, under the pseudonym "Hippocrates," to the columns of "BUILDING."

It has been the author's aim to give an account of the usual condition in which plumbing work, done years ago—and some done quite recently—may be found, and also to give suggestions on the proper manner of doing the work. The title "Hints" has been chosen purposely, for this little volume cannot, and does not pretend to be, an exhaustive *treatise* on the subject.

Frequent reference has been made to the Report on "Filth-Diseases and their Prevention," by John Simon, Chief Medical Officer of the Privy Council and of the Local Government Board of Great Britain. Several quotations from this important essay have been taken as standard truths, upon which much of the subject matter has been founded.

Other writers on Dwelling-house Sanitation have been quoted. The author's object in doing so was to strengthen his own assertions by the conclusions, reached by other workers in the same field.

Doubtless there will be those, who, on perusal of the book, will find "nothing new in it." To these, the author would reply that, in his judgment, the subject of "Healthy Homes" cannot be too often brought to the attention of the public, who, as a rule, are yet indifferent to the importance of having in their houses *good drainage and sewerage*

WM. PAUL GERHARD.

New York, January 1st, 1884.

CONTENTS.

CHAPTER IV.

Traps and Systems of Trapping.

CHAPTER V.

Details of Traps.

CHAPTER VI.

Insecurity of the Common Water Seal Traps.

CHAPTER VII.

Defects in the Plumbing Work of Dwellings.

CHAPTER VIII.

Cellar Drains and Drainage of Cellars.

CHAPTER IX.

Usual Defects of House Drains; Sewer Connections, Privy Vaults and Cesspools.

CHAPTER X.

System of Internal Sewerage as it should be in a Dwelling.

CHAPTER XI.

Plumbing Fixtures.

CHAPTER XII

Removal and Disposal of Household Wastes.

HINTS
ON THE
DRAINAGE AND SEWERAGE OF DWELLINGS.

CHAPTER I.
FRESH AIR VERSUS SEWER GAS.

THE subject of this little volume is one to which, at some time or other, every one engaged in building must necessarily devote some attention. Architects, engineers, builders, mechanics, physicians and sanitarians, house-owners and house-holders, are all interested in it. While it is only one branch of the problem of house sanitation, the sewerage of the dwelling is of more than ordinary importance, as from it largely depends the attainment of the conditions described in the old Greek sanitarian's formula: "*Pure air, pure water and a pure soil.*"

In planning a new house, its site and location, the character of the subsoil of the building lot, the aspect of the house, the construction of proper foundations and dry, well-lighted cellars, the means for preventing dampness of walls, the proper materials for building, the arrangement of rooms, halls, closets and staircases most consistent with health, comfort and convenience, the lighting, warming and ventilation of the house, its drainage, water supply, sewerage, the arrangement of plumbing fixtures and plumbing work, the removal and proper disposal of kitchen garbage, slops, ashes, of excreta and liquid wastes of the household, and many other details, must be carefully considered.

The *soil* on which the house will be erected should be free from impurities, and must be constantly kept unpolluted; an abundance of fresh *air* of proper temperature and a continuous removal of vitiated air are necessary for the health of the inmates; a never-ceasing and bountiful supply of pure and wholesome *water* is required for drinking and cooking purposes, for daily ablutions of the body, for cleaning utensils, washing linen, scrubbing floors, windows, flushing plumbing fixtures, etc.

The water brought into the dwelling under pressure must be removed from it after use, being then more or less foul and mixed with the discharges of the human body, from soiled linen, from personal ablutions, with greasy matters of the pantry and kitchen, etc. Such fouled water from the household (to which may be added the foul liquids from stables and manufacturing establishments of all descriptions) is called *sewage*, and the object of a *sewerage* system is the immediate removal by means of water of all sewage from habitations, and its disposal in a manner so as to render it not only innocuous, but, if possible, useful.

For convenience in performing the various duties of domestic cleanliness, and further for health and comfort's sake, our modern houses are furnished with set fixtures, basins, tubs, sinks and water closets, supplied with hot or cold water, and each connected by wastepipes to the drainage system. The planning of a proper and efficient system of water supply and sewerage for a dwelling requires a thorough knowledge of the subject, which is of such a vast extent that it seems impossible to offer here more than a few hints and suggestions. These will relate not so much to proper mechanical execution of the details of plumbing work as they will to sound sanitary arrangement of the drainage system. Upon the latter will largely depend the future immunity of the building from sewer gas, and consequently the freedom of its inmates from certain preventible diseases, generally attributed to the entrance of gases from the sewer or drain through defective plumbing work. It is not my purpose to discuss at length the much vexed question of the influence of sewer air in developing or spreading certain epidemic diseases. I believe that this question can only be satisfactorily solved by physiologists, and that neither architects, nor engineers nor physicians should pronounce an opinion of their own, unless they should have devoted years of actual study and experimenting to organic chemistry and to that branch of physiology which relates to the germ theory of disease.

Dr. John Simon, Chief Medical Officer of the Privy Council and of the Local Government Board of Great Britain, and a high sanitary authority, says in his able report on "Filth Diseases": "An important suggestion of modern science with regard to the nature of the operations by which Filth, attacking the human body, is able to disorder or destroy it, is: that the chief morbific agencies in Filth are other than those chemically-identified stinking gaseous products of organic decomposition which force themselves on popular attention. Exposure to the sufficiently concentrated fumes of organic decomposition (as, for instance, in an unventilated old cesspool or long-blocked sewer) may, no doubt, prove immediately fatal by reason of some large quantity of sulphide of ammonium, or other like poisonous and fetid gas, which the sufferer suddenly inhales, and far smaller doses of these fetid gases, as breathed with extreme dilution in ordinary stinking atmospheres, both give immediate headache and general discomfort to sensitive persons temporarily exposed to them, and also appear to keep in a somewhat vaguely depressed state of health many who habitually breathe them ; but here, so far as we yet know, is the end of the potency of those stinking gases. While, however, thus far there is only the familiar case of the so-called *common chemical poison*, which hurts by instant action, and in direct proportion to its palpable and ponderable dose, the other and far wider possibilities of mischief which we recognize in Filth are such as apparently must be attributed to *morbific ferments* or *contagia ;* matters which not only are not gaseous, but, on the contrary, so far as we know them, seem to have their essence, or an inseparable part of it, in certain solid elements which the microscope discovers in them : in living organisms, namely, which in their largest sizes are but very minute microscopical objects, and at their least sizes are probably unseen even with the microscope; organisms which, in virtue of their vitality, are indefinitely self-multiplying within their respective spheres of operation, and which therefore, as in contrast with common poisons, can develop indefinitely large ulterior effects from first doses which are indefinitely small. Of ferments thus characterized, the apparently essential factors of specific chemical processes, at least one sort—the ordinary septic (putrefactive) ferment—seems always to be present where putrefactive changes are in progress, as, of course, in all decaying animal refuse; while others, though certainly not essential to all such putridity, are in different degrees apt, and some of them little less than certain, to be frequent incidents of our ordinary refuse. As, apparently, it is by these various agencies (essential and incidental) that Filth produces "zymotic" (fermentative) disease, it is important not to confound them with the fetid gases of organic decomposition; and the question, what infecting powers are prevalent in given atmosphere, should never be regarded as a mere question of stink. It is of the utmost practical importance to recognize in regard of Filth, that agents which destroy its stink may yet leave all its main powers of disease-production undiminished. Whether the ferments of disease, if they could be isolated in sufficient quantity, would prove themselves in any point odorous, is a point on which no guess need be hazarded; but it is certain that in doses in which they can fatally infect the human body they are infinitely out of reach of even the most cultivated sense of smell, and that this sense (though its positive warnings are of indispensable sanitary service) is not able, except by indirect and quite insufficient perceptions, to warn us against risks of morbid infection."

Abundant evidence has been given by Dr. Parkes in his "Manual of Practical Hygiene" to establish with almost absolute certainty the fact that there is a connection between sewer air and certain preventive diseases, notably bowel diseases.

Notwithstanding this, Dr. Soyka and Dr. Renk, both of Munich, have denied the existence of any positive proof of the influence of sewer gas upon the spread of zymotic diseases.*

Their views are in direct conflict with the above quoted theories of Dr. Simon, with the overwhelming evidence given by Dr. Parkes and other writers, and the facts and statements contained in many Health Reports of this and of the old country.

While this scientific question is pending, it seems best to continue to assume that gases originating from the decomposition of animal or vegetable matter, especially if the decomposition goes on in the absence of oxygen, are capable of doing harm, when entering a dwelling. Just how much harm they may do will largely depend upon the constitution of the individual exposed to the influence of such germ-containing atmosphere. A healthy person, having much out-of-door exercise, may breathe sewer air with impunity; on the other hand, people in delicate health, women and children, may suffer severely from breathing impure

* See Deutsche Viertel jahrschrift für oeffentliche Gesundheitspflege, Vol. XIII. and XIV., 1881-1882.

air, the consequences being slight headache, nausea, vomiting, or diarrhœa, dysentery, enteric fever, cholera, diphtheria, etc. Workingmen engaged for a whole day in cleaning sewers may feel less influence of the deadly poison than a person sleeping in an unventilated room containing an untrapped washbowl or other plumbing fixture.

It has been said that "pure air and plenty of it is the best cure for sewer gas." This is undoubtedly true, but how little is it yet understood! *Pure air* is just what is needed in our homes, and I shall repeatedly refer in the following pages to the necessity of an abundant supply of this life-giving element in order to effect the proper ventilation of all living and sleeping rooms, in particular of all closets and bathrooms of a dwelling, of all plumbing fixtures, soil and waste pipes, of the house drain, the sewer or cesspool.

Says Dr. George Wilson in his book, "Healthy Life and Healthy Homes": "In order to keep the air of the house pure and healthy, there must be no damp foundations, no damp walls, no dark and dingy cupboards or corners to confine the air and devitalize it, no filth in or around the dwelling to pollute it, and no overcrowding. There should be cleanliness everywhere, adequate means of ventilation, plenty of window space to let sufficient light into every room and proper appliances for warming during cold weather." And Dr. John Simon speaks about domestic cleanliness as follows: "The perfection of cleanliness would be that all refuse matters should from their very beginning pass away inoffensively and continuously;

and the principles of approximation to that ideal must evidently be, first to provide to the largest practicable extent for the continuous outflow of refuse as fast as produced, and secondly (so far as continuous outflow cannot be got) to provide for the closest possible limitation and the completest possible innocuousness of such refuse as is unavoidably detained."

This last quotation embodies in the fewest words the vital principles of household sanitation.

Let us now inquire what the actual condition of city houses is, with reference to those rooms containing plumbing fixtures.

CHAPTER II.

NECESSITY OF VENTILATION IN ROOMS CONTAINING MODERN CONVENIENCES; DEFECTIVE ARRANGEMENT OF PLUMBING FIXTURES.

BATHROOMS and water closet apartments are frequently located in the center of the house, with no other light but that from a window opening into the staircase hall. One looks in vain for any means of renewing the air of the apartment. In placing the bathroom in this part of the house, it

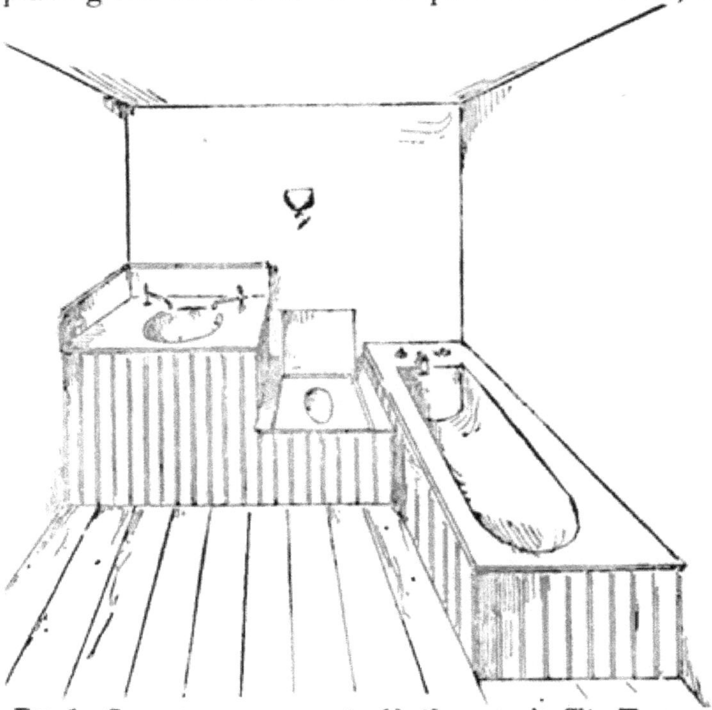

FIG. 1.—Common arrangement of bath rooms in City Houses.

certainly did not occur to either architect, house owner or plumber that, just in cases where a room containing plumbing work cannot have a window to an outside wall, ventilation is most essential. more so than any costly furniture, decorated ceilings or artistic wall paper. It is not unusual to find water closets and urinals placed in dark closets, lighted by a gas flame (Fig. 1), with no other outlet for the products of combustion, and any possible foul gases, than into the hall of the house, or, what would be infinitely worse, into an adjoining bedroom. Is it surprising, then, that complaints of sewer gas are frequent and loud?

On the bedroom floors we find a dark, damp, unventilated and ill smelling closet, which contains a slop-sink or a slop-hopper, into which the housemaid pours the slops from bed-chambers. Such a closet is certainly as much in need of ventilation as the water closet apartment is, for slop-sinks have large surfaces exposed to spatterings, and, as usually constructed, receive no flush of clean water following a discharge from a slop pail; they remain fouled with dirty matter, which soon gives off offensive odors. But, in ninety-nine out of every hundred houses, the air of such a closet is never changed, except when its door is opened, and then only to bring its fouled atmosphere in connection with the air of the very centre of the house.

Descending into the basement, we find in many cases a nuisance created by the servants' water closet. The most remote, ill-lighted and closely confined corner of the basement or cellar is generally selected for it. Is there anything astonishing about the usual condition in which we find such apparatus? The closet being located in a dark, out-of-the-way place, no trouble is taken in keeping the bowl free from filth. I have seen, in the houses of wealthy, refined and intelligent people, such places in the very worst state of neglect and untidiness, being seldom, if ever, looked after by heads of families. Such condition of things warrants the general conclusion that the occupants of

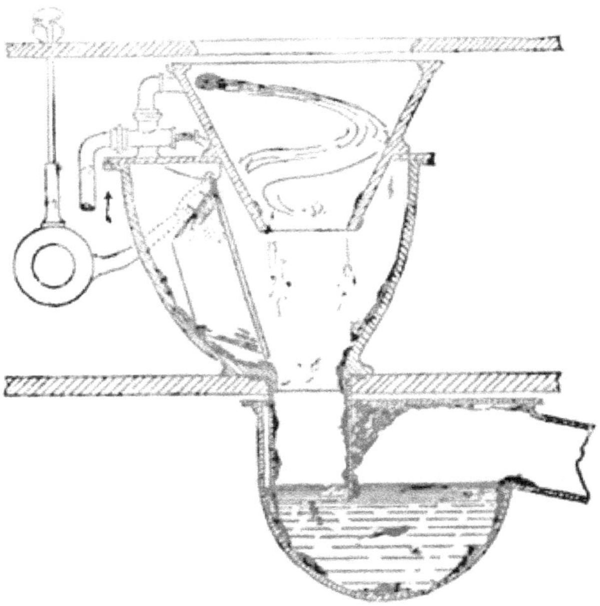

FIG. 2.—Sectional view of Pan Closet.

such houses do not at all consider that filthiness of the servants' water closet apparatus is not confined to its apartment, but may create gases of decomposition that will spread out and fill the whole dwelling, to say nothing of unclean habits to which it must lead servants.

I fear that many readers will call my description

exaggerated—others may think the account rather discouraging—but my picture is by no means overdrawn. I know from long actual experience that the facts disclosed are the rule, not the exception, in most of our houses. To arouse public interest in this question, and to enlighten those in search of a healthy home, such facts and statements should have the widest publicity given them.

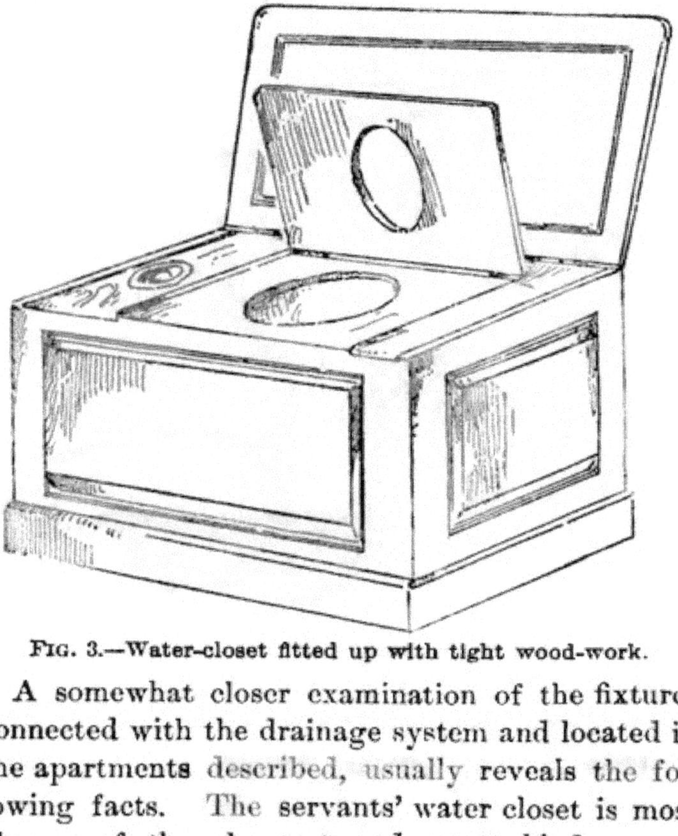

FIG. 3.—Water-closet fitted up with tight wood-work.

A somewhat closer examination of the fixtures connected with the drainage system and located in the apartments described, usually reveals the following facts. The servants' water closet is most always of the cheapest and worst kind, a *pan closet* (Fig. 2), encased in *tight nailed woodwork* (Fig. 3), with no ventilation under the seat nor to the apartment. If we succeed after considerable trouble and delay in removing the riser, seat and cover, we will find the floor stained from leakage of the closet valve, and ill-smelling from the absorption into the wood of spilled urine. Dust and dirt and perhaps vermin will have accumulated in the hidden corners. The closet bowl is generally flushed by a valve, supplied directly from the rising water main of the house. The flushing water enters the bowl at one point of its circumference (Fig. 2) and whirls around, unable to flush the bowl, which accounts for its generally filthy appearance. The operation of pulling the handle starts the flush, at the same time it causes the pan which closes the outlet of the bowl to tilt, thus dumping its contents into the container. Each time this is done, a puff of sewer gas from the container enters the apartment. This container or receiver has been called a "hidden chamber of horrors." As usually constructed it is of plain iron, with rough interior surface, of large size to allow the movement of the pan, and receives no flush whatever. Its sides soon get coated with excrements, putrefaction begins, and sewer gas is thus generated in the heart of the house. The plumber may have assured the house owner that he has put a trap under the closet to cut off the gases from the soil pipe, he may have told him that there is an additional seal against gases afforded by the water in the pan, and seeing all the complicated machinery about the apparatus, (see Fig. 4,) the householder—generally a layman in such matters—will be led to believe that he has in his house the most modern and perfect appliance.

And a most perfect and ingenious apparatus it is—to fill the house with noxious, nasty and health-menacing smells! For there are ways in which the sewer gas will enter the room, even if the pan should be closed and the closet outlet sealed. The hole in the container for the spindle which works the pan is never made tight, thus establishing a direct connection between the container and the room.

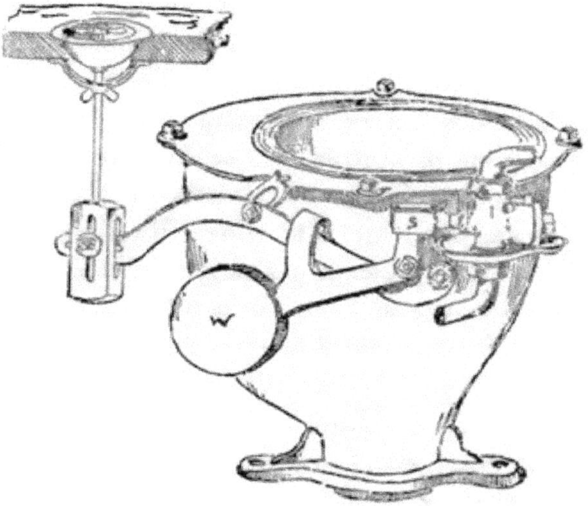

FIG. 4.—Outside view of Pan Closet, bowl removed.

The bowl is fastened to the container only by a putty joint, which crumbles away in time, or is eaten by rats, thus opening another road for the gases of the container. The trap of the water closet is another source of annoyance; it must necessarily accumulate excremental matter (Fig. 2.), as the valve flush is not sufficiently strong to drive such matter through the dip of the trap. In old houses this trap is often of the worst kind, a D-trap (Fig. 5), which in a short time becomes a filthy cess-

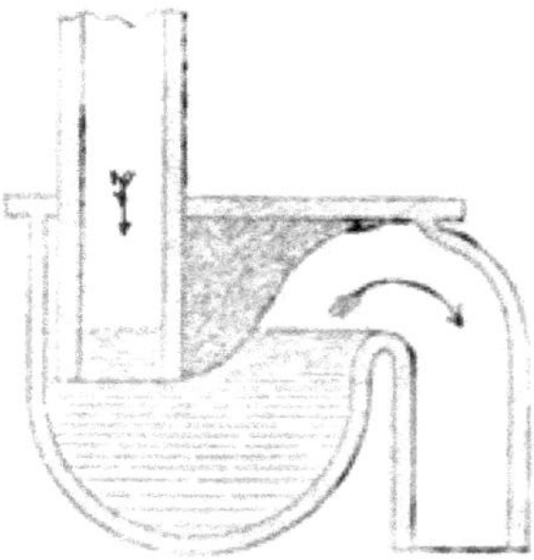

FIG. 5.—D-Trap.

pool in the room. The pan is quickly corroded by the action of sewer gas from the container, and thus the security of the double waterseal is lost, and the bowl loses its water and becomes more readily fouled on this account. The floor under the pan closet may be provided with a safe to catch drippings, and its waste pipe is in many cases run into this trap, below its water line (Fig. 6). Thus the foul water from the trap standing back in the drip pipe will evaporate into the apartment.

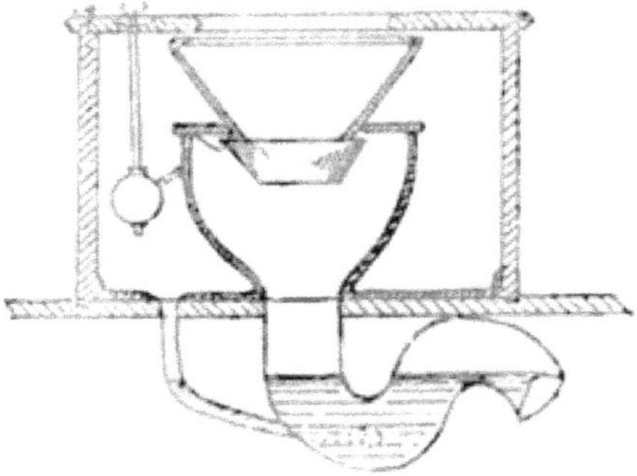

FIG. 6.—Drain pipe from safe under water closet delivers into the water closet trap.

In some instances a cheap kind of hopper, generally of iron, is used for servants' closets, and is no less objectionable than the pan closet, its flush being mostly insufficient to keep the rough inside of the hopper free from excreta.

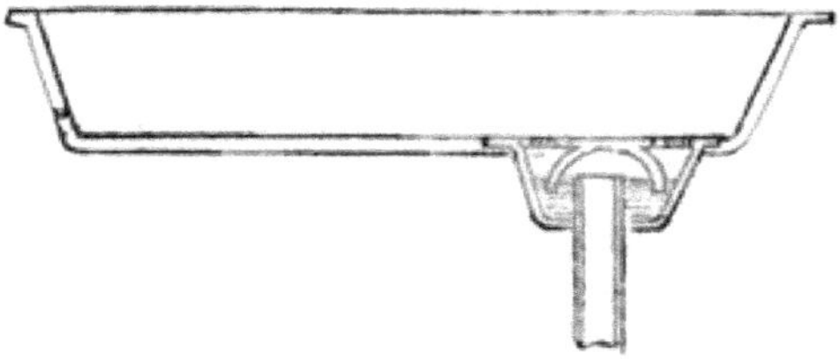

FIG. 7.—Sectional view of Kitchen Sink, with Bell Trap and Strainer.

The *kitchen* is always provided with a *sink*, usually of iron, but sometimes of soapstone or other material, connected to the nearest soil or waste pipe by a branch pipe of lead. This latter is, as a rule, too large in diameter, and consequently accumulates deposits. It frequently joins the soil pipe or the main drain without even the interposition of a trap. In more recently built houses the outlet of the sink is trapped, but this trap is either faulty in design, for instance a bell trap, or, if an S-trap, it is much too large and consequently ill-flushed. Our illustration (Fig. 7), shows a sink with bell trap, forming in its upper part a strainer. As the latter is removable

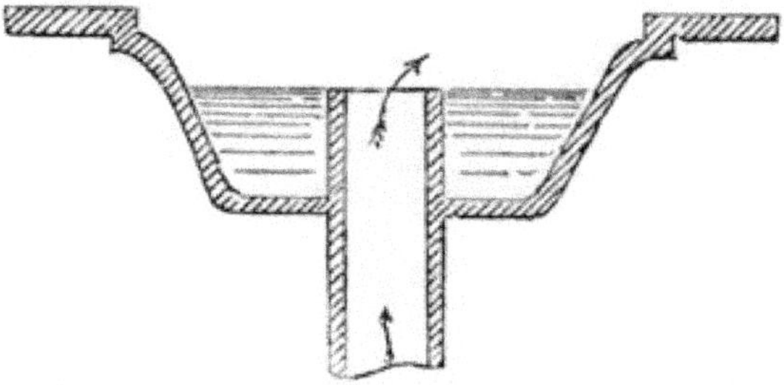

FIG. 8. —Bell Trap with Strainer removed.

in most houses, kitchen servants readily acquire the pernicious habit of lifting it to brush all kind of refuse into the outlet. They hereby not only cause frequent obstructions of the trap and waste pipe, but they establish a direct connection between the kitchen and the gases of the sink waste pipe, as long as the strainer is removed from its place (Fig. 8). In older houses I almost always find the kitchen sink encased with carpentry, (Fig. 9.)—such is often the case even in our modern homes—and the foul, dark space underneath the sink is utilized for the storage of cooking utensils, kerosene cans, cleaning rags, old shoes, scrubbing brushes and other matters. (Fig. 10.)

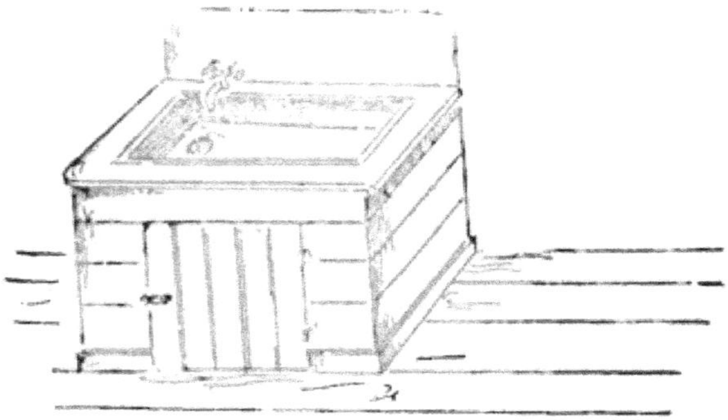

FIG. 9.—Kitchen Sink fitted with wood-work.

Laundry-tubs in the laundry-room or kitchen are most always of wood, and contribute, after long use, their share to the pollution of the air in the house. Wooden tubs are objectionable, not only because they get leaky, but because wood absorbs the f .th of soiled linen, and is difficult to clean;

FIG. 10.—Filthy condition of interior space underneath kitchen sink.

consequently, wooden tubs, when old, give off a very offensive odor. Moreover, they are generally closed up tightly underneath, and the floor will be ill smelling from leakage and will rot in time. The trapping of laundry wastes is also frequently defective.

The *copper sink* in the *butler's pantry* off the dining-room is seldom, if ever, free from defects. By removing the boards encasing the space under the sink, we will find a large reservoir or bottle trap on the large waste pipe (Fig. 11). Ample size of both was probably deemed necessary by the plumber who did the work, to prevent the choking up of the pipe and trap with grease. But the small stream from the pantry sink has not been able thoroughly to flush the waste pipe, and the bottle trap, placed to act as a grease trap, which therefore should have been frequently cleaned, was forgotten, not being easily accessible, and left to take care of itself. It is now almost filled with putrid grease, the sides of the waste pipe are coated with a similar matter, the overflow pipe from sink forms also a channel for gases, and thus has the air in the butler's pantry been continually contaminated.

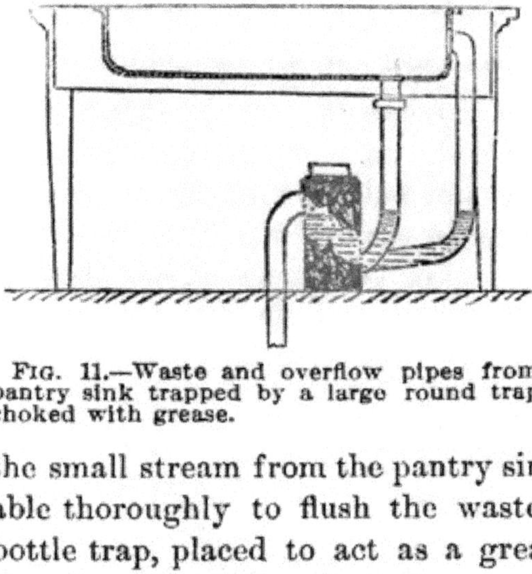

FIG. 11.—Waste and overflow pipes from pantry sink trapped by a large round trap choked with grease.

The toilet-rooms and the bath-rooms contain further fixtures, which the plumber and carpenter took particular pains to enclose with tight carpentry (Fig. 11). The water closet, although generally a pan closet flushed from a valve, is in some cases supplied with water from a special cistern, or else it is of a more expensive pattern, either a *valve*

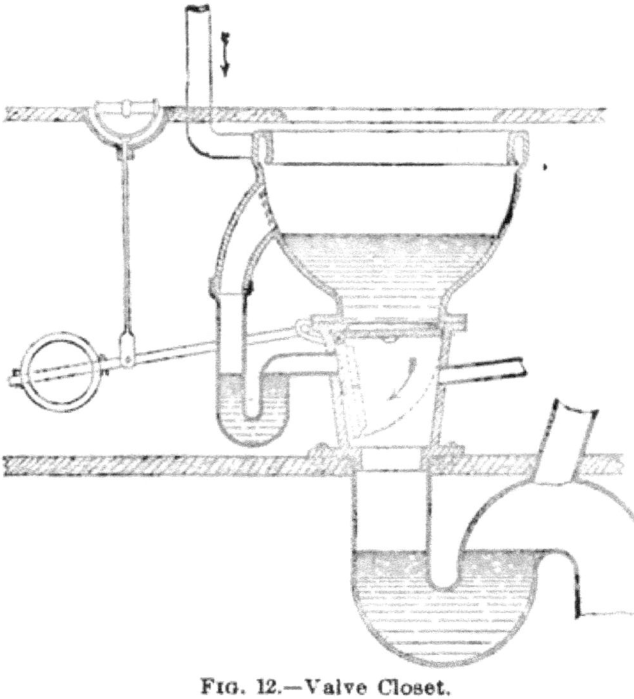

FIG. 12.—Valve Closet.

closet (Fig. 12) or a *plunger closet* (Fig. 13). But either of these has serious defects, and although both are great improvements upon the pan type, neither will in the end prove satisfactory. The valve of valve closets leaks after long use, consequently the water will run out of the bowl. Should the use of closet be continued while in such a condition, the flap valve will become coated with filth, as will also the walls of the container. The same coating with filth occurs in time with the plunger and plunger chamber of the second type of closet mentioned. In both cases decomposition of organic matter will go on within the walls of the house, which decomposition it is the object of a proper system of house drainage to prevent by an immediate removal of all waste matter.

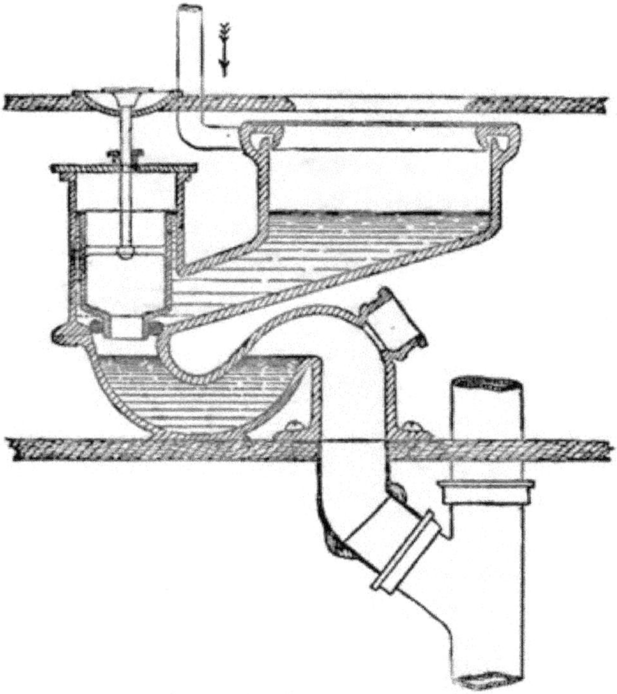

FIG. 13.—Plunger Closet.

The wash bowls and *bath tubs*, as well as their traps and waste pipes, are seldom properly designed and constructed. Their wastes are often left in direct communication with a soil or waste pipe (Fig. 14), in other cases they are trapped only by running them into the water closet trap below its water line (Fig. 15). If this trap be displaced or its contents siphoned out, a free communication is established between the soil pipe and the bath room. If the wastes enter the water closet trap above the water line, the gases of the container will find a ready exit at the bath or bowl. A common defect of bowls and bath tubs consists in their overflow pipe joining the waste pipe beyond its trap (Fig. 16). And even where overflow and waste pipe are both trapped by the common S-trap (Fig. 17), the water in the latter may be removed by siphonage, or it may evaporate, should the bowl not be used for some time (Fig. 18). Where the soil and waste pipes, into which the bowl or bath tub wastes deliver, have no ventilation by being extended through the roof, it may happen that the S-trap is forced by back-pressure, and also that the water in the trap absorbs gases and possibly germs of disease, which may be given off on the house side of the trap, that is into the room, when the water in the trap is agitated.

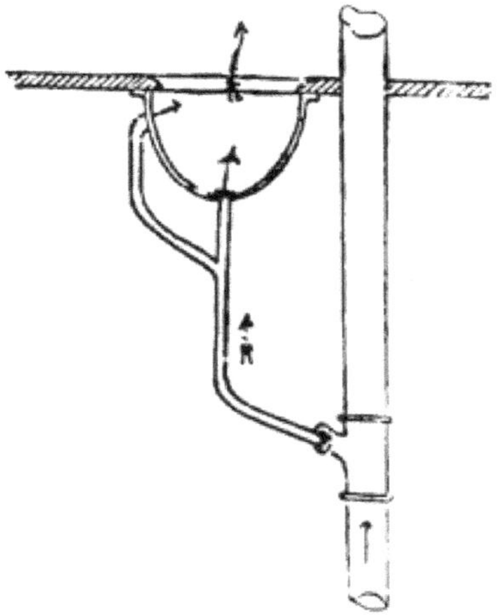

FIG. 14.—Waste and overflow pipes from wash bowl in direct communication with soil pipes.

Stationary wash stands have either common bowls with outlet at the bottom, closed by a plug, hung to a brass or plated safety chain, or else they are of the "tip-up" type, in which case the bowl is emptied by tilting its contents into a larger concentric bowl underneath. Both arrangements are apt to become filthy; soapsuds remain sticking to the many links of the chain, which is difficult to clean, and the lower bowl of a tip-up basin presents generally, upon investigation, a far from satisfactory appearance. Being covered and not easily accessible, it is never cleaned, and filth gradually accumulates in it, and its putrefaction may soon cause great annoyance. The chain and plug arrangement for bath tub is in no respect better than that for wash bowls.

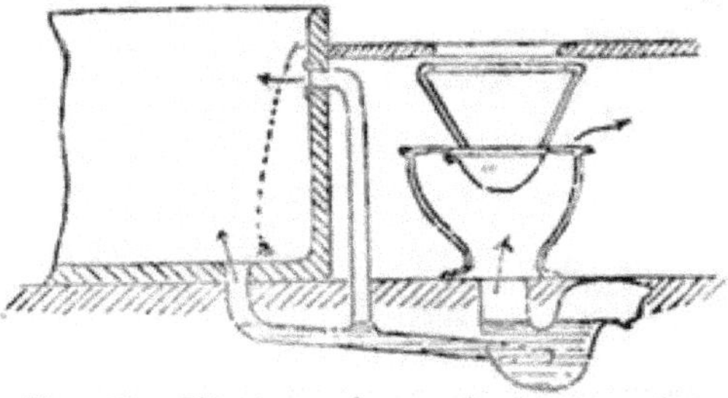

FIG. 15.—Waste and overflow pipes from bath trapped by running into water closet trap below its water line.

Bad smells from wash bowls or bath tubs are occasionally traced to the overflow pipe of either kind of fixture. The walls of such pipes remain coated, should an occasional overflow occur, with slime, and receive no flushing whatever. Generally a very long length of these pipes remains in communication with the

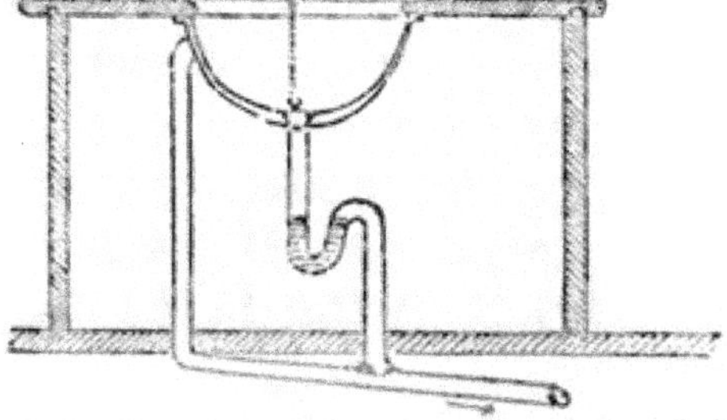

FIG. 16.—Waste from wash bowl trapped from overflow pipe joining waste beyond trap.

room, continually to foul its air. For these reasons overflow pipes should be dispensed with wherever possible; and that such arrangement can often, though not always, be had, will be shown later.

Stationary bowls and tubs, when lined with a safe to prevent damage to ceilings, have drip pipes to carry off overflowing water in case of accidents. The arrangement of such drip pipes is frequently deficient. They are sometimes in direct connection with soil or waste pipes; in other cases they are trapped, and the traps become ineffective by evaporation of the water. In houses of a more recent construction a weeping pipe is arranged to supply water to such trap at frequent intervals, but even such a device is unsatisfactory and dangerous.

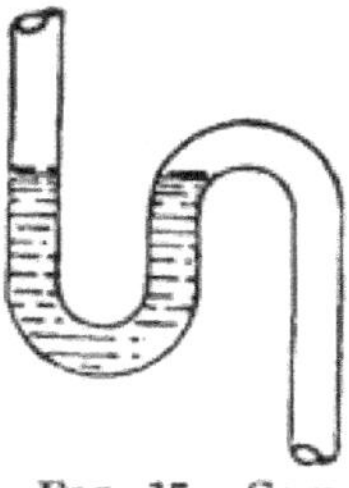

FIG. 17.—Common S-trap.

A fixture common to office rooms and to lavatories or toilet rooms adjoining billiard rooms in private houses is the *urinal*. It is usually in an extremely nasty condition, and its appearance most unsightly, owing to the feeble flush from a stop-cock, which is unable to cleanse the urinal. Urine remains spattered on the bowl and is sometimes spilled on the floor, and its rapid decomposition creates most pungent and disgusting odors. Unless of an approved pattern, with plenty of water in the bowl, and with a strong flush of water driven through a flushing rim and derived from a cistern, I should hesitate to recommend a

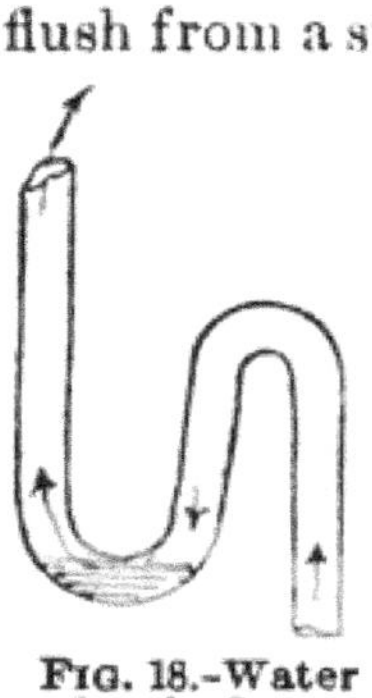

FIG. 18.-Water seal of S-trap lost by syphonage or evaporation.

urinal for a private house. The improved water closets, of which I shall speak hereafter, can generally be so constructed and put up as to be used in place of a urinal.

Our remarks have thus far been confined to the usual arrangement of rooms containing plumbing fixtures, and to the condition in which such apparatus is frequently discovered to be upon examination of houses built some years ago. Just here the fact should be mentioned that in cities where plumbing work is now regulated by law and controlled by plumbing inspectors of the Board of Health, a marked influence upon the quality and general character of the plumber's work may easily be recognized. It is somewhat sad to think that, to secure a good, safe job from the *average* plumber, regulations as to the details of his work had to be drawn up by the city health authorities, and that constant vigilance is necessary to secure the proper carrying out of these regulations. The general public, in particular the vast number of families in cities who are dependent for shelter on tenement houses, apartments, or small houses built for speculation and for rent, should be thankful for the many benefits derived from the enforcement of such plumbing regulations. Briefly stated, bath rooms in the center of the house are now always located near light and air shafts. Less wood work is used in fitting up sinks and tubs ; water closets and wash bowls have hinged doors to render possible frequent inspections of hidden parts of fixtures. The pan closet, although still extensively used, is supplied from special flushing cisterns, and has a properly vented S-trap. In addition to this, every tub, bowl, sink, etc., is provided with a separate vented trap.

The next chapter will treat of soil and waste pipes and their usual defects.

CHAPTER III.

SOIL AND WASTE PIPE SYSTEMS AS USUALLY FOUND IN DWELLINGS.

FAULTY arrangement of soil and waste pipes in dwellings aggravates the danger arising from defective plumbing fixtures. The principal defects to be considered are: improper material for pipes, bad manner of making pipe joints, insufficient or defective ventilation of the soil and waste pipe system, and use of pipes of too large calibre.

I shall not dwell upon the well-known defects of lead soil pipes; although still the rule in England, they have, fortunately, in this country; become a thing of the past, and when found upon examination of houses built years ago are invariably condemned and removed. As the origin of lead soil pipes dates back from the time when ventilation of waste pipes was not yet practised, they are found corroded and honey-combed by the action of sewer gas (See Fig. 19). Cast iron pipes with socket joints have since then taken their place. They are sold in lengths of five feet, with a single or double hub, and innumerable fittings are manufactured to provide for changes of direction, for branch wastes, etc. In Fig. 20, *a.* represents the single and double hub pipe, *b.* is a Y branch, *c.* is a double half-Y branch, *d.* is a quarter bend with double hubs, *e.* an eighth bend, *f.* a sixth bend, *g.* an increaser, *h.* a T-branch, *i.* an offset, and *k.* a long quarter bend.

FIG. 19.—Lead Soil Pipe, unventilated and corroded by sewer gas.

In ordinary contract work, the plumber always uses what is called "light soil pipe," often not even protected against rust by a coating of coal tar pitch, and therefore a very flimsy article of manufacture, which should not be tolerated wherever sound work is expected. This pipe is condemned by Mr. James C. Bayles in his book "House Drainage and Water Service" in the following words: "An iron soil pipe should not be too light. In much of the cheap work of the time the pipe used is lighter than it should be. I have seen pipe set up in houses which, tested with callipers, I have found to be not more than one-eighth of an inch thick. The objections to this kind of pipe are numerous and important. It does not possess the requisite strength, it is too quickly eaten through by rust, and it is very apt to have sand holes and other imperfections, which, for a time, may afford an easy outlet for the gases of the sewer. The difference in cost between light pipe and that of suitable thickness (a quarter of an inch for private houses and three-eighths and upward where there is a long line of large size to accommodate a continuous outflow of considerable volume) is not great enough to make the economy profitable."

"In architects' specifications we seldom find a suitable weight of iron pipe called for. Consequently the principal demand is for very cheap and light pipes. As made, they are as hard as chilled iron—owing to the fact that they are cast so thin—and about as brittle and difficult to cut as glass. If dropped they crack or break, and *are utterly untrustworthy at all times.*" (The italics are mine).

The *Sanitary Engineer*, in commenting upon the lately passed Boston plumbing ordinance, says: "We regret, however, that all lines of soil pipe, and the fittings on it, over fifty feet in length, were not required to be the standard extra heavy pipe, which is one-fourth inch thick; the pipe specified in the law is too thin for the high buildings now being erected. The additional cost of the extra heavy pipe is only for the value of the iron, the labor being the same, and for reasons often stated in these columns, and which every plumber knows, it should not be for a moment considered when the risks to be avoided are taken into account."

The better grades of soil pipe, the *heavy* and *extra heavy* soil pipe and fittings in the market, the price of which is about double that of light pipe, are usually specified only for public or other large expensive buildings.

My experience with extra heavy cast iron soil pipe warrants me in saying that even the latter is very often decidedly bad, having an uneven thickness of metal, and consequently being in its

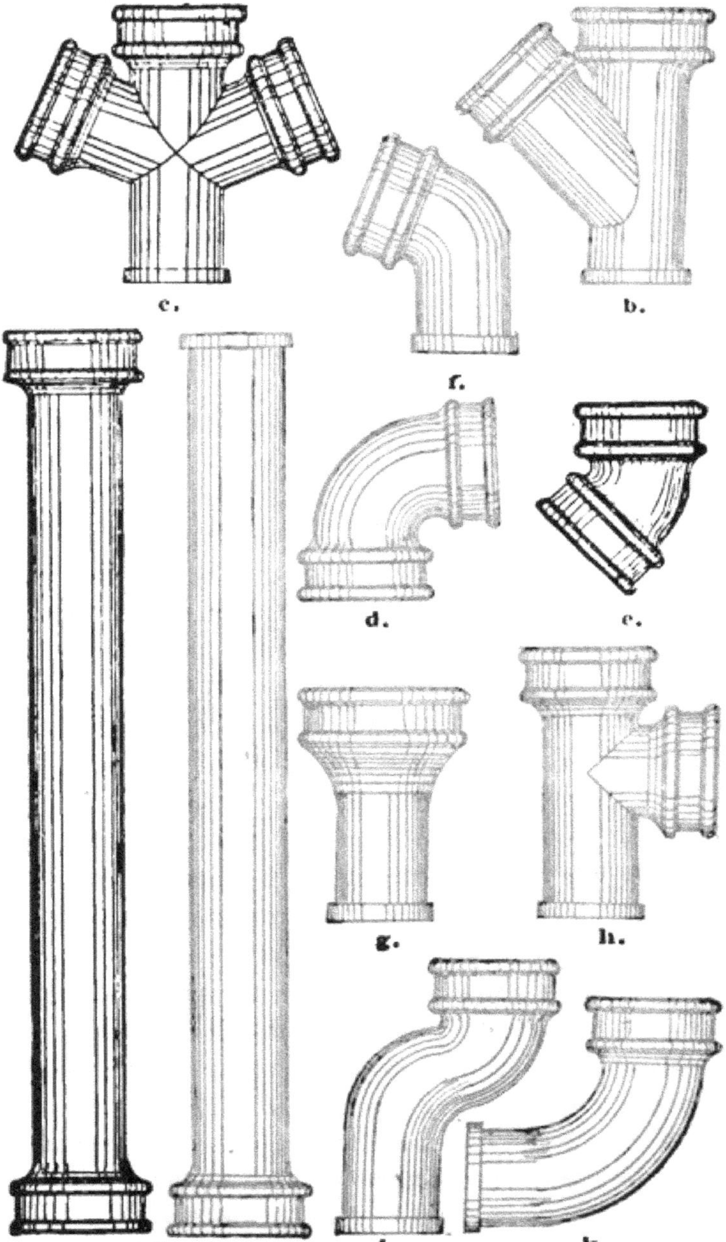

FIG. 20.—Cast Iron Plumbers' Pipe and Fittings.

weakest part no thicker than "light" pipe. As in all other engineering structures, the strength and durability of a system of drainage should be determined by the strength of its weakest point, and thus it will be readily understood that extra heavy soil pipe is no better, although more costly, than light pipe, as long as the manufacturer takes no pains to secure a *uniform* thickness of the metal. Plumbers' soil pipe is never tested at the works, and sand holes or flaws are a common occurrence, and are not readily detected by subsequent inspection, especially if the pipe is coated with tar or asphalt, or enamelled. An equally weak point with plumbers' pipe is the shape and strength of the hub, as from it depends the tightness of the joints. One of the worst defects in plumbing work of cheaply built houses is the manner of tightening the joints in cast-iron soil pipe.

No other part of a common plumbing job shows so many defects as a stack of iron soil or waste pipe; there is scarcely another detail in a system of drain pipes for a dwelling in which so much rascality or criminal stupidity is shown than in the manner of making joints in iron pipe, and this is especially the case wherever architects or builders tolerate such pipes to be built into walls, inasmuch as under such circumstances defective joints are readily covered up and brought out of sight. Such pipes are often jointed with paper, covered with sand, or else some cheap mortar is thrown into the space between spigot and socket; in other cases putty is used, or red lead. Wherever joints are in sight some lead is perhaps poured on top of the sand to give the joint the appearance of having been done with the proper material. Other workmen are content with filling the joint with lead poured in hot, omitting the most important operation, that of caulking the joints after the lead has cooled off. But even where a gasket of hemp or oakum, a ladle full of hot lead and caulking tools are used, carelessness or ignorance of the mechanic have much to do with improper and leaky joints. The manner of applying the gaskets of oakum, the quality of the melted lead, its purity, the temperature to which it is kept in the pot on the fire, the manner of pouring the lead, and finally the operation of caulking it after shrinking, these are all details worthy of careful consideration, but unluckily, seldom looked after in plumbing a dwelling.

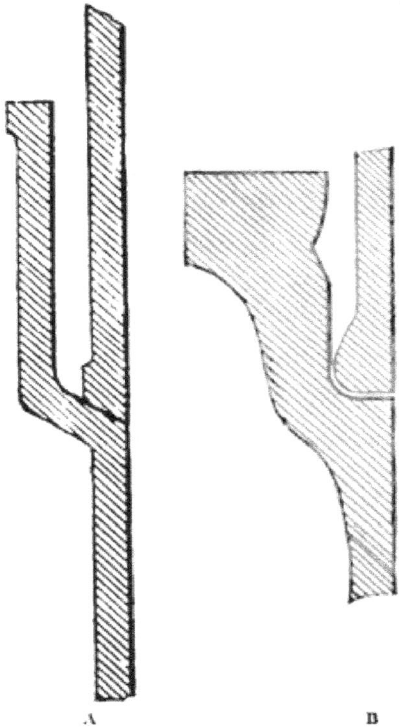

FIG. 21.—Sketch showing comparative strength of hubs of plumbers' pipe and gas pipe.
A—Plumbers' pipe. B.—Gas pipe.

It would not require much reflection on the part of the mechanic to know that the safety of the occupants of a house must depend to a great extent upon the perfect tightness of all joints in waste pipes. Unfortunately, however, the health of the inmates is not a matter usually considered by the speculative builder or the average plumber. Joints in cast-iron soil pipe could be made tight,—if the thickness of the pipe hubs would be increased, if the pipe would be carefully selected, inspected, and tested with hydraulic pressure before leaving the foundry, or at any rate before coating the pipes with a rust preventing solution,—but even then it would require proper care and a good deal of attention in making the joints. Unless the Board of Health regulations require the testing of all soil pipes in new buildings under the supervision of inspectors appointed by the Board, or unless an expert engineer superintends the drainage work in a dwelling, care is seldom, if ever, taken to attain such results. If the subsequent testing of soil and waste pipes shows a leakage, the plumber is very apt to excuse himself by throwing all blame

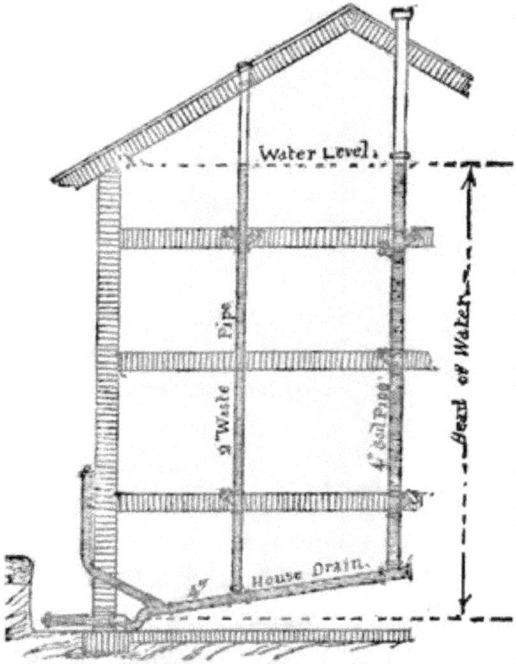

FIG. 22.—Sketch showing method of applying the water pressure test.

upon the manufacturer. He will claim, and I have frequently heard the statement made, that the latter does not manufacture pipes with hubs of sufficient strength to withstand the severe knocking occasioned by the caulking tool. This, I admit, is true, but it cannot be considered a valid reason for not making tight joints. I believe that if plumbers would join in the earnest protest of the best architects and civil engineers against such "light soil pipe," or against extra heavy pipe with uneven thickness of metal or hubs of insufficient strength, they would be able to secure a better article of manufacture.

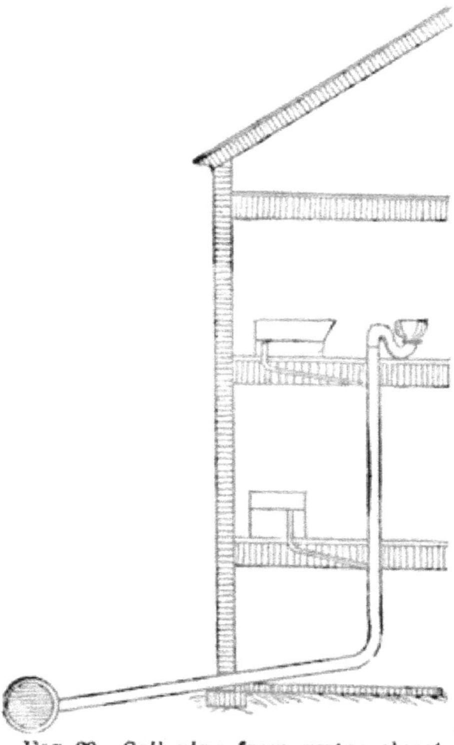

FIG. 23.—Soil pipe from water closet trap to the sewer without extension through the roof. No fresh air pipe and no trap on main drain.

It has been my personal observation that honest and conscientious plumbers—with best possible intentions to do only first-class work—were frequently unable to caulk the lead of joints sufficiently tight without splitting the hub of the pipe. In other cases the joint could not be made tight owing to the impossibility of reaching all parts of the lead in a joint with the usual caulking tools, the soil pipe being located in a recess or a partition.

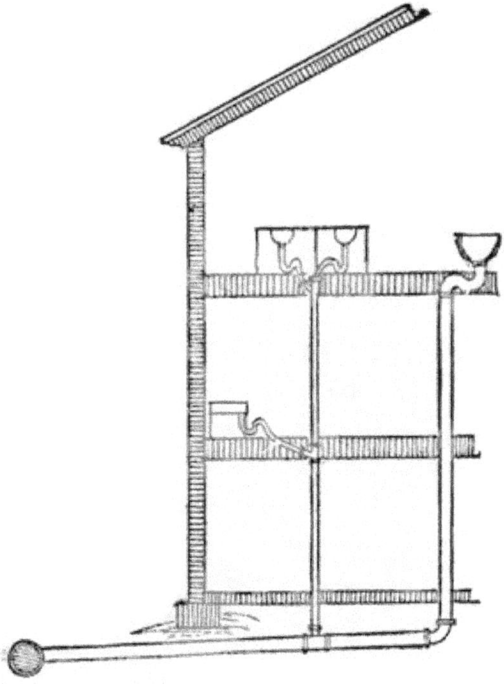

FIG. 24.—Soil pipes and waste pipes without ventilation.

That cast-iron socket pipes *can* be tightly jointed will be at once apparent by referring to gas and water mains. In the one case leakage of illuminating gas, and in the other waste of water, can be effectually prevented by properly made joints. A comparison with water pipes would, perhaps, be considered unfair, as these are expected to stand much heavier inside pressures than a soil pipe of most houses, when tested by water pressure. The pressure in gas mains is, however, very slight, seldom exceeding one or two pounds per square inch. Let us then compare the common plumbers' pipe with the pipe used for conveying illuminating gas. Fig. 21 shows a cross section through the bells of heavy plumbers' pipe and of gas pipe. It will be readily admitted that the hub B is designed to resist strong knocking by the mechanic's tool, while the hub A is apparently weak and therefore frequently broken.

Tightness of joints may easily be tested and defects in the piping detected by the "water pressure test." Before setting and joining any fixtures to the soil and waste pipes, all its outlets are closed by india rubber plugs, squeezed with iron discs by means of a bolt and nut, and the pipes filled with water, Fig. 22. Should there be a leak it is readily detected and should be immediately remedied. The test is then repeated until there are no more signs of a leak. For very high buildings, for instance the flats now being erected in many parts of New York city, the head of water would become too great, and in such case the pipes are tested in sections. This test is undoubtedly more useful than the peppermint or smoke test; it is easily applied, and is one of the most important

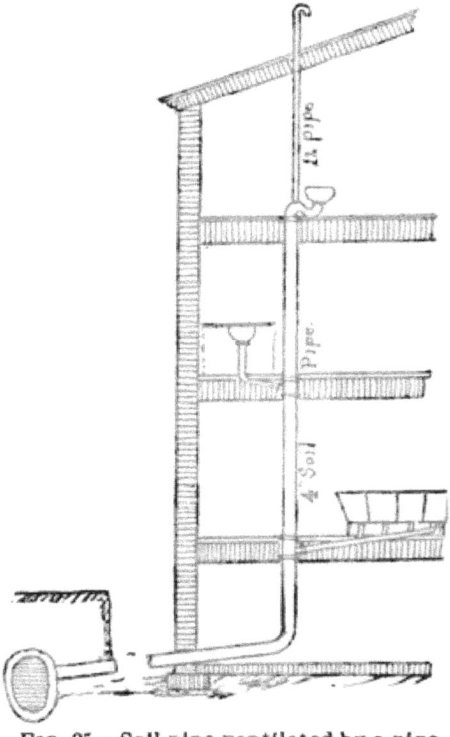

FIG. 25.—Soil pipe ventilated by a pipe of insufficient size, extended up to the roof.

things in connection with the plumbing of dwellings. It is very desirable that it should be more frequently applied by the Plumbing Inspector of the Health Board in cities where plumbing is regulated by law than is done now; for a house with a network of waste pipes, that have successfully stood this test, is a much safer place for human beings than most houses of the present day.

Under the heading, "Testing Soil and Waste Pipes by Pressure," the editor of the "Metalworker" has recently given his opinion as follows:

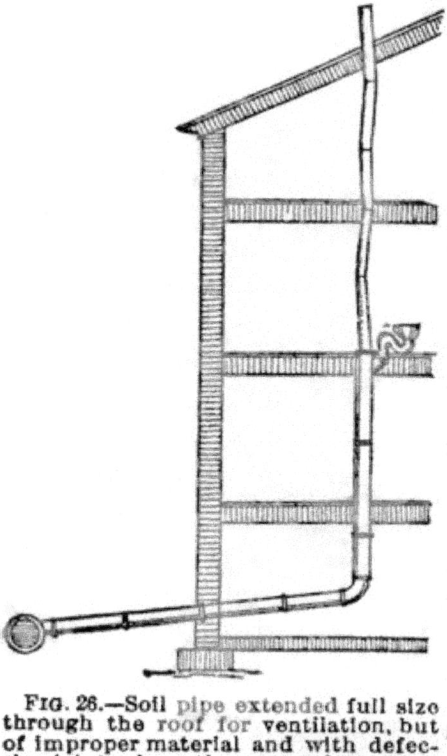

FIG. 26.—Soil pipe extended full size through the roof for ventilation, but of improper material and with defective joints above the water closet trap.

"We suppose that no one who has had occasion to inspect plumbing work in houses already completed, has not many a time felt a strong desire for some means which should enable him to determine whether a given line of pipe was sound or leaky. In our own experience recently, we met a case of this kind, in which it was almost impossible by ordinary methods to make ourselves certain in regard to the condition of several lines of soil and waste pipe. The houses were so arranged that the soil pipes cannot be opened for inspection through their whole lengths, and even the most careful peppermint test will not give all the information that is desirable. A pipe may be tight and apparently sound, yet of so thin a substance that the least pressure will destroy it or break it through. Joints may be tight at the moment, though barely filled with a thin coating of putty, blown out almost at a single breath. Such pipes, though tight for the moment, are not safe against the slightest pressure, and at any time may be liable to have their continuity broken by a slight jar."

"*The longer we study this subject the more completely do we become convinced that the true remedy for this state of things is a test of the soil pipes by pressure. Scamping is so easily done and so difficult of detection that it seems impossible to avoid it, even in the best jobs which may be constructed. A large proportion of the work is done in difficult situations, where the workman has every temptation to save himself labor and discomfort, and in such situations poor work is the rule rather than the exception.*" (The italics are mine). * * * * * "The real objection to such a test is to be found in the fact that it calls for perfect workmanship throughout. It demands just what every house builder and house owner wishes to have, but just what it is very difficult to obtain from even the best plumbing establishments in the city. In gas fitting, which is much less difficult than plumbing work, no sane man would dare to trust a large job without carefully testing it under pressure. We do not think that it will be many years before the method of testing by pressure

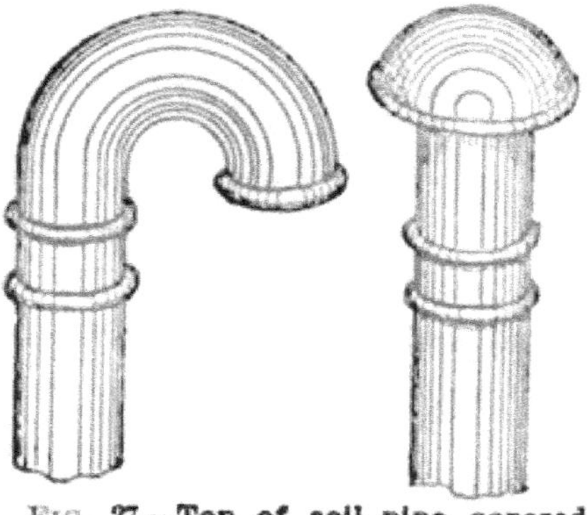

FIG. 27.—Top of soil pipe covered with return bend or ventilating cap.

will be made a requirement in the best jobs of plumbing work."

In a previous chapter, lack of ventilation was recognized as a serious defect of plumbing fixtures and their apartments. Not less serious is the insufficient ventilation of soil and waste pipes. There are still thousands of houses in every large city where soil pipes have no air circulation whatever, but stop at the trap of the highest water closet, and where waste pipes are run only from the drain in the cellar to the fixtures, such as sinks, tubs, bowls, etc., without upward extension (See Figs. 23 and 24). In many cases the plumber thinks that he has provided a sufficient ventilation by running a small (1½ or 2 inch) vent pipe through the roof (See Fig. 25). In a few cases only is the extension of the soil pipe of the full size of the pipe.

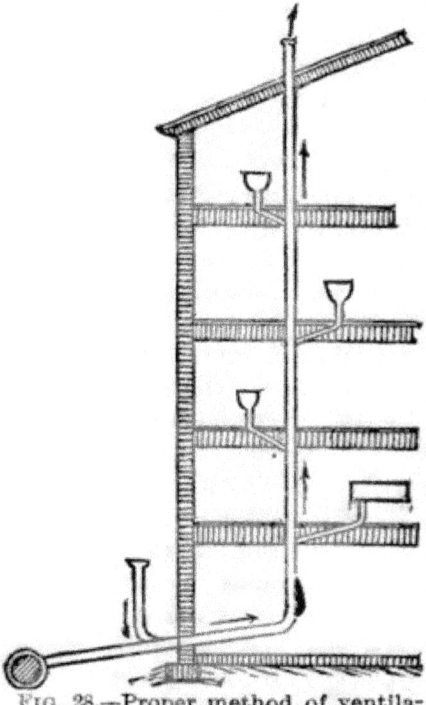

FIG. 28.—Proper method of ventilating a soil pipe.

That this ventilating extension should be of the same material of which soil pipes are made is a rule which is often violated by skin plumbers. Galvanized iron or tin pipes are frequently run from the highest water closet upward through the roof, the joints being imperfectly closed or not made at all, the pipes being simply slipped one into another (See Fig. 26). To illustrate, I quote from a late issue of the "Metalworker:"

"A few days ago a friend in the trade called my attention to one of the most startling instances of rascally plumbing of which I have ever heard. A friend of his, living in Brooklyn, was troubled with bad odors in his house and sickness in his family. He was advised to determine whether his soil pipe was tight, and the use of peppermint was suggested. Some was procured, and the householder went to the roof for the purpose of pouring it down the soil pipe through the projecting extension. His wife and others were stationed at different points in the house to see if they could detect the smell of peppermint at any of the fixtures. As they were unable to do so, it seemed as if the test had failed to show any defects in the pipe system, but in a few minutes all the oil of peppermint which had been poured down the pipe came through the ceiling of one of the bedrooms on the highest floor, and dripped down upon the carpet. Examination revealed the fact that the supposed ventilating extension of the soil pipe above the roof was a mere sham. The plumber had put on a length of pipe and secured it in an upright position, but it had no connection whatever with the soil pipe of the house. In the four houses immediately adjoining, all of which were built under the same contract, the same condition of affairs was found."

Even where the extension of the soil pipe is of proper size and material, its object is often defeated by a ventilating cover or hood or return bend placed on top of mouth of pipe, which greatly impedes ventilation (Fig. 27).

All these attempts at establishing an air current are futile, unless a second opening for fresh air is provided at the foot of the iron soil pipe. With two openings of the full size of soil pipe a constant current and dilution of the air in the pipe, and a destruction of organic matter coating the inner walls of pipes is effected (See Fig. 28).

It is a mistake to place any ventilator over the mouth of soil or waste pipes. While some cowls may act very efficiently with certain directions of the wind, it is now believed that for the usual direction of the wind a plain open-mouthed tube affords greatest upward movement in vertical pipes.

Great carelessness is often shown in the location of the fresh air pipe as well as of the soil pipe mouth. The former should be remote from windows, and the latter not too near any skylight, air shaft or chimney top.

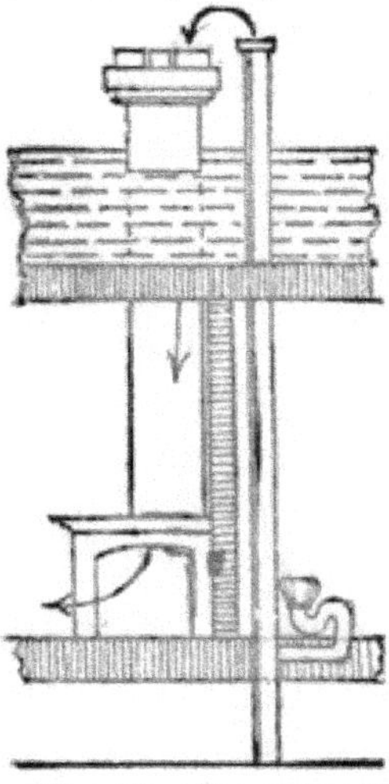

FIG. 29.—Top of soil pipe located too near a chimney top.

Fig. 29 illustrates how sewer gas may be carried down a chimney flue and enter the dwelling through fire places, if proper care is not taken to locate the soil pipe mouth remote from and at least a few feet below chimney tops; down drafts in chimney flues or ventilating shafts are known to occur at times, and may thus be the cause of annoying gases in rooms.

Fig. 30 shows a soil pipe terminating above the roof close to a mansard roof window, perhaps of an attic dormitory. The injudiciousness of such location is quite apparent.

FIG. 30.—A soil pipe terminating near attic window.

Experience has also clearly demonstrated the need of enlarging the extension of smaller waste pipes to four inches diameter (Fig. 31), for smaller openings above the roof become frequently obstructed in

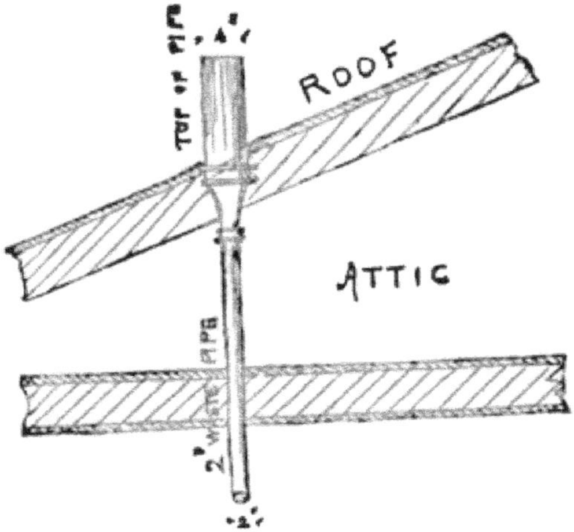

FIG. 31.—Waste pipe enlarged at roof.

cold climates by hoar frost, and thus the purpose of the pipe extension is practically annihilated.

It is a common mistake with plumbers and builders to make the soil and waste pipes unnecessarily large. Soil pipes of 5 or even 6 inches diameter are used where a 4-inch pipe would be ample to carry off all the waste water that could be discharged into it. Such a pipe is sufficient for dozens of water closets on the same or on different floors. In my own practice I never use a soil pipe larger than 4 inches diameter, and where only one water closet has to be served I should not hesitate to use a 3-inch pipe, provided I could rely upon a judicious use of the closet and upon constant use of the now universal toilet paper, and provided also the traps on waste pipes connected to the 3-inch soil pipe are efficiently protected against siphonage. Where, on the other hand, water closets are subjected to rough treatment and are made the receptacles of all sorts of rubbish, not properly belonging thereto, I fail to see the wisdom of using a larger pipe, say of 5 or 6 inches bore, as under such conditions obstructions are just as likely to happen with large as with small pipes. The right remedy would seem to me to be to teach people the proper and judicious use of such fixtures. Where vertical waste pipes are required to receive the water from sinks, bowls and tubs, located at a distance from the soil pipe, experience has proven a 2-inch pipe sufficiently large; pipes of larger sizes will always remain imperfectly flushed, and therefore become, in time, extremely foul.

Mr. Hellyer discusses this question in his "Lectures on the Science and Art of Sanitary Plumbing," under the heading of "Size of Soil pipes," as follows:

"About fifteen or twenty years ago it was common for plumbers to fix (under the direction of a specification) a 6 inch soil pipe when it had to take the branches of four or five water closets, and with many architects and builders, as well as the plumbers, of to-day, 5 inch and 4½ inch are the general sizes, and that too for only one water closet.

Now, as it is of the utmost importance that a soil pipe should be efficiently flushed out with water every time a water closet upon it is used, it is evident that the smaller the size of the pipe the more efficiently will it be flushed, and as it is not wanted for a coal-shoot or a dust-shaft, I cannot see why it should be so much larger than the *outlet-way* of the water closet into it. In private houses, where the water closets would be used with greater care than in public buildings, I consider 3½ inch lead soil pipe* quite large enough to take a *tier* of three or four water closets. I am supposing the soil pipe to be ventilated at top and bottom, and each trap or branch ventilated as well. I consider 4 inch soil pipe, when of lead, and made by hydraulic pressure, large enough to take the branches from several more water closets, and 4½ inch soil pipe is ample to take a tier of six or seven double closets, fixed over each other in a seven-storied building, for though many of them might be used together they would not be discharged precisely at the same moment of time, and one or two seconds would suffice, in a vertical soil pipe, for the discharges to keep clear of each other, and if they did mingle it would not so much matter, so long as the traps, by efficient ventilation, were made proof against any disturbance that could take place by the simultaneous use of all the closets upon the piping. I have had 3½ inch lead soil pipes fixed to the

---

*Mr. Hellyer, following the usual English custom, prefers lead as material for soil pipes.

tiers of three and four water closets, but have never known the smallest inconvenience from such an arrangement, while the pipes, as far as I have been able to see, have kept cleaner than 4½ inch soil pipes near them—i. e., under the same conditions.

In public buildings—as warehouses, hotels, banking houses, stations, club-houses, etc.—the soil pipe ought, perhaps, to be larger—say 4 inch, but I consider 4 inch (or 4½ inch) large enough for any place and for any number of closets.

In many places, with efficient water service, 3 inch soil pipes might be fixed for single water closets without any risk of stoppage."

Mr. Hellyer also mentions that he fixed in his factory for the use of the workmen a 3 inch soil pipe, with branches for three water closets, a Hellyer Valve Closet, a Hellyer Vortex and a Hellyer Artisan Hopper Closet, and the size is found to be quite large enough, though the closets are rarely ever idle during the working hours.

The effect of the official supervision of plumbing regarding soil and waste pipes shows itself more in the improved ventilation than in the material and jointing of such pipes. Frequent inspections of the work, especially if it is done by contract, will render impossible cases of scamping, such as the following, related by the editor of the "Metalworker":

"When work has to be given to the lowest bidder, without regard to his honesty or responsibility, carelessness in the drawing of specifications is attended with very serious danger. In a house which I know of, the fact that the soil pipe was merely carried about a foot under the cellar floor and left open, with absolutely no connection with the sewer, was not discovered until the insufferable smell in the cellar revealed the fact that all the foulness which had been discharged from the house since it was first occupied had accumulated in the pipe, and soaked into the earth of the cellar bottom. The plumber claimed that the specification on which he made his bid did not call for a sewer connection. Of course he knew that it ought to have been made, but when required to bid below the cost of honest work, he considered himself perfectly at liberty to take advantage of any omission in the specifications."

CHAPTER IV.

TRAPS AND SYSTEMS OF TRAPPING.

BY extending all soil and waste pipes at least full size through the roof, and providing an inlet for fresh air on the line of the house drain, we have established a circulation of air through the waste pipe system. (Fig. 32). The system shown in the sketch is, however, still imperfect. Although it is a common occurrence to find waste pipes of dwelling houses thus arranged, some further provisions are required to render the system complete. No amount of ventilation would suffice to keep the air pure in houses having a drainage system arranged on the plan shown in Fig. 32. Sewer air would penetrate them from cellar to attic, saturating bedding, upholstery, carpets, furniture, wall papers, causing loss of strength and health of the occupants, and frequently breeding disease, or even causing the death of some beloved member of the household.

The reason why such an arrangement of the pipes is defective is quite obvious. Should the house drain deliver into a cesspool or connect to a sewer in the street, it affords, in both cases, a chance for escape of generally very foul gases into the house pipes. But in addition to such gases from the sewer or cesspool, the soil and waste pipes of every house contain more or less foul air (improperly called "*sewer*" gas), derived from decomposing waste matters adhering to and coating the inside of the waste pipes. With the arrange-

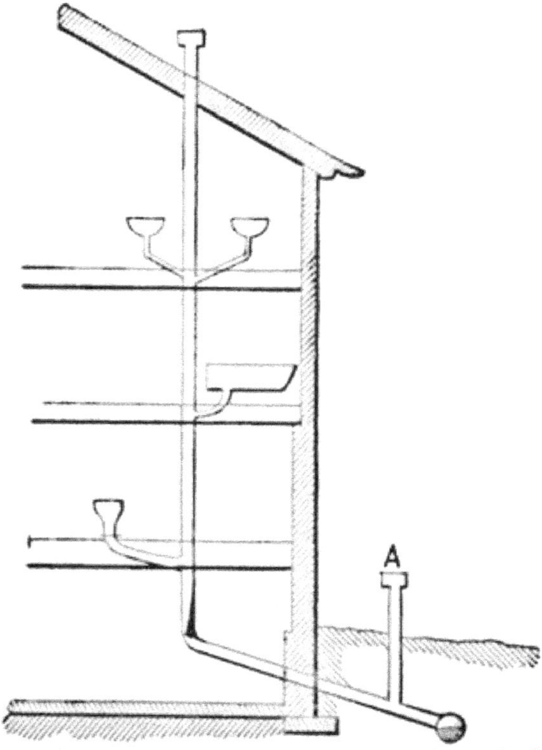

FIG 32 —Soil pipe extended full size through roof; fresh air inlet at foot, but no traps under fixtures or on the main drain.

ment shown, soil pipe air, as well as cesspool or sewer gases, would find a ready outlet through the branch waste pipes and fixtures into the room. To prevent this some barrier ought to be placed on waste pipes and drains, which allows the foul water to run off, at the same time making it impossible for gases to return through such channels. This is what is commonly called "*trapping*" a drain or waste pipe, and the following remarks will be chiefly devoted to *traps.* The simplest trap is a bend in the pipe (Fig. 33), retaining sufficient water to "form a seal." It must be admitted that every trap is, to a certain extent, an obstruction to the free flow of water, and brings with it the danger of occurrence of deposits and consequent decomposition of organic waste matter, but in *a system of house drainage traps are necessary evils.*

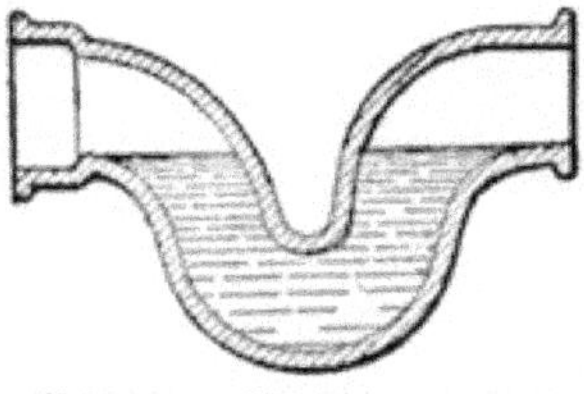
FIG. 33.—Siphon, or running trap.

First in importance is the proper trapping of all fixtures of a dwelling. Each water closet, urinal, slop sink, wash bowl, bath tub, sink and set of laundry tubs should be separately trapped as near to the fixture as possible by a reliable trap. Improperly trapped or untrapped fixtures are fully as much, if not more so, the cause of bad and unhealthy odors in dwellings as improper and defective joints in soil pipes.

If, then, we put a trap under each and every plumbing appliance (Fig. 34), it still remains our duty to prevent any escape of foul gases of the sewer or cesspool, into the soil and waste pipes, or at the opening A, which is intended to act as an inlet for fresh air. Waste pipes, as we have seen above, always contain more or less foul air, which should be diluted and rendered harmless as much as possible by introducing into the pipes a constant current of *pure* air. A trap should, therefore, be placed on the line of the house drain, between the fresh air pipe A and the sewer or cesspool. (See

Fig. 35). The opening at A will now almost continually act as an inlet, except when a discharge through a soil pipe occurs, at which time the current may, for a short time, be reversed. As long

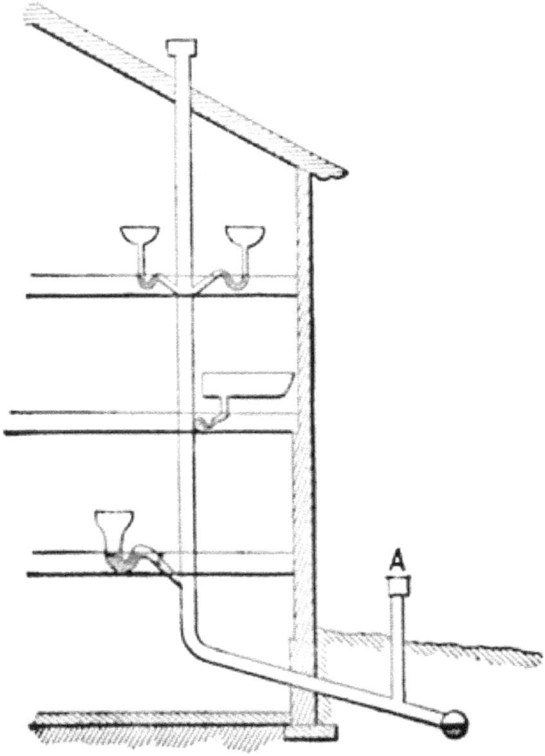

FIG. 34.—Soil pipe extended full size, and provided with foot ventilation; each fixture in the house trapped by a trap, but no trap on the main drain.

as such inlet is judiciously located, remote from windows or piazzas, or the cold air box of the heating apparatus, a downward current through the soil pipe is unobjectionable.

Much diversity of opinion exists in regard to the necessity of trapping the main drain and the fixtures.

There are experienced men who claim that the fresh air pipe A and the trap on the main drain should be

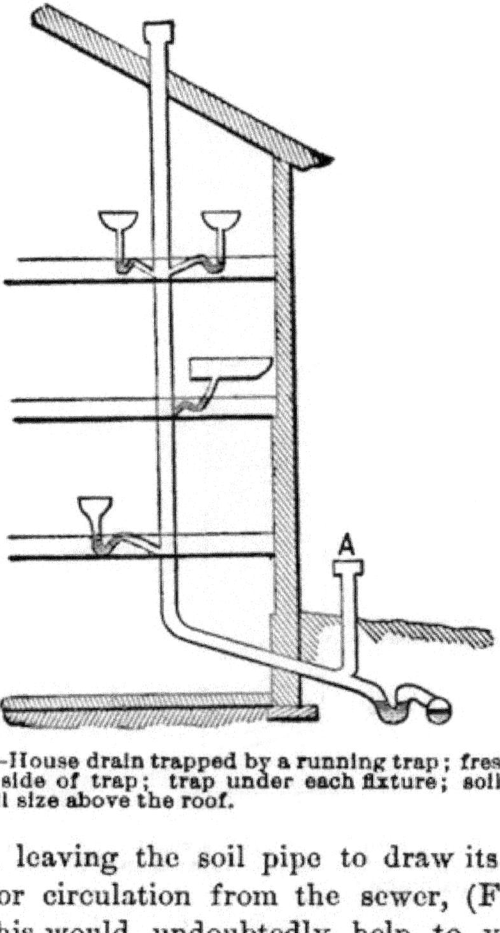

FIG. 35.—House drain trapped by a running trap; fresh air pipe on house side of trap; trap under each fixture; soil pipe extended full size above the roof.

omitted, leaving the soil pipe to draw its supply of air for circulation from the sewer, (Fig. 36). While this would undoubtedly help to ventilate the sewer, I have sincere doubts as to the wisdom of a more general application of such a system. I should certainly condemn it severely wherever a house drain discharges into a cesspool, which is always more or less foul. It would also be wrong, in my opinion, wherever the street sewer is known to be so foul as to constitute an "elongated cesspool." With *well jointed pipes* and *well-trapped fixtures* it may be possible to allow *well constructed* and

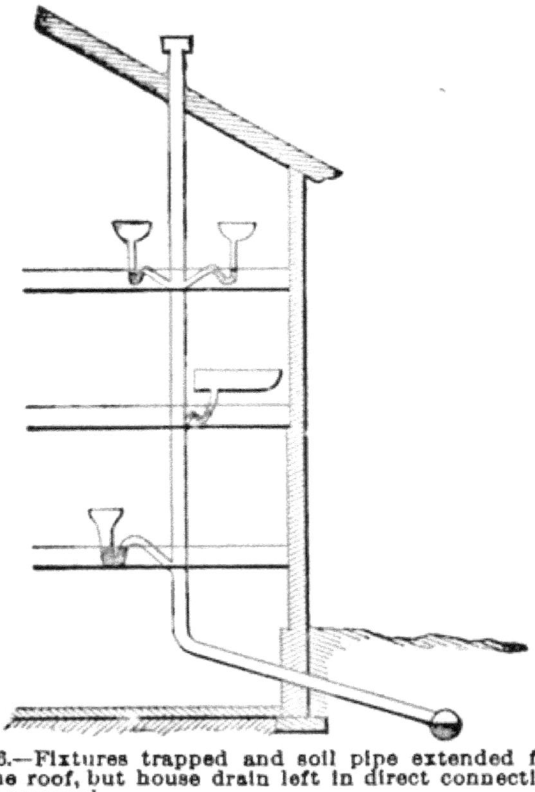

FIG. 36.—Fixtures trapped and soil pipe extended full size above the roof, but house drain left in direct connection with sewer or cesspool.

copiously flushed sewers to breathe through the house pipes, but up to the present day such work, as regards both the drainage arrangements of dwellings and the construction of sewers, has been the exception rather than the rule.

Where a house drain of a single house empties into a river or lake or into the sea, and the distance from the house to the outlet is moderate, the trap on the main drain may be omitted, always suppos-

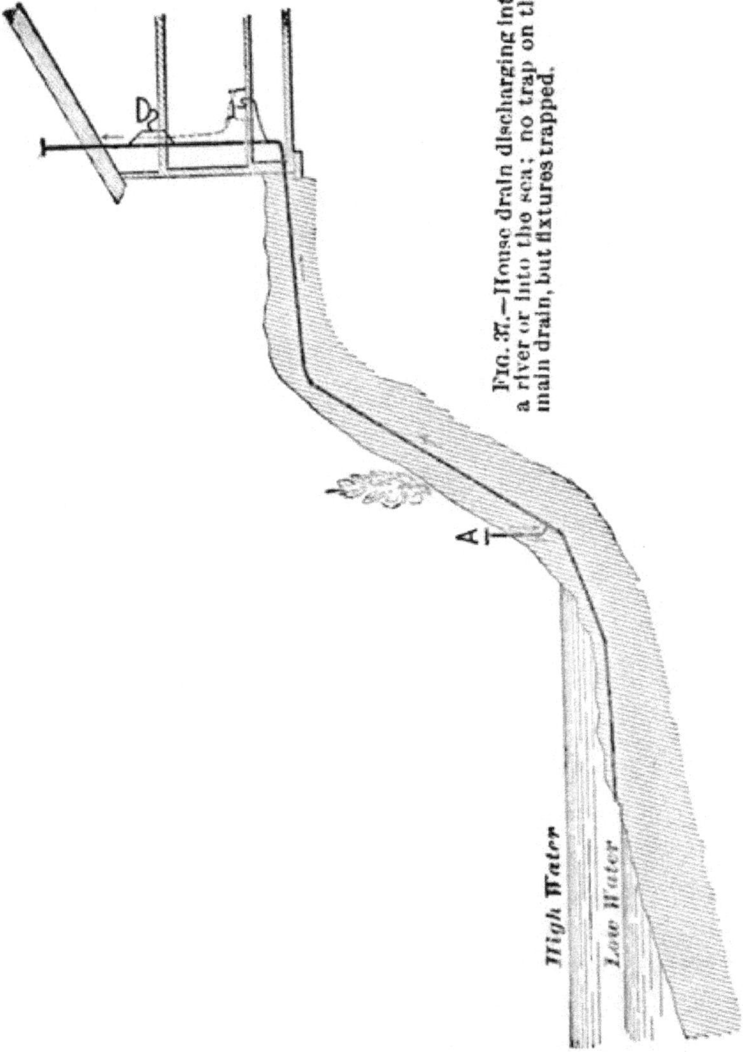

FIG. X.—House drain discharging into a river or into the sea; no trap on the main drain, but fixtures trapped.

ing the work in the house to be done in the most approved and perfect manner, to be thoroughly inspected from time to time, and the drain to be of proper material, laid with ample fall, and securely and tightly jointed. Should the outlet be located so as to be closed at times, by high tide or otherwise, it is necessary to construct a fresh air inlet A, entering the drain just above the highest possible water level. (Fig. 37).

In the majority of cases however, my decided preference is for "disconnection" or complete isolation of each dwelling from the cesspool or the common sewer. Mr. Mansergh, a civil engineer of large experience, ably discusses this question as follows:

"I would detach as far as is practicable every house from the main sewer. As a part of a whole sewerage system, every single house is brought more or less closely in connection with every other house, and by this means evils existing in some houses may become common to all. The more perfectly this connection can be severed the better. The aim in all cases should be to isolate as far as possible, but at all events to cut off the direct communication to the interior."

The "Model Bye-Laws of the Local Government Board of England" require a *suitable trap to be placed in every main drain of a building*, and add the following explanatory note:

"The object of this clause is to prevent foul air, as from public sewers, from making its way into house drains. Public sewers ought to be ventilated otherwise than through house drains, the more so as it is in the power of householders to ensure the efficiency of their own drains, but they are unable to control faulty construction leading to deposit, etc., in public sewers. It is also only by the adoption of such a clause that houses can be protected against the influence of infectious matters received into the common sewers. In a similar way buildings should be protected against foul air from cesspools when such means of drainage outfall have to be adopted."

It has also repeatedly been proposed to leave out the traps under fixtures, sometimes substituting for the traps a downward draft through the fixtures by connecting them with a heated flue. The advocates of this system (Fig. 38) require, of course,

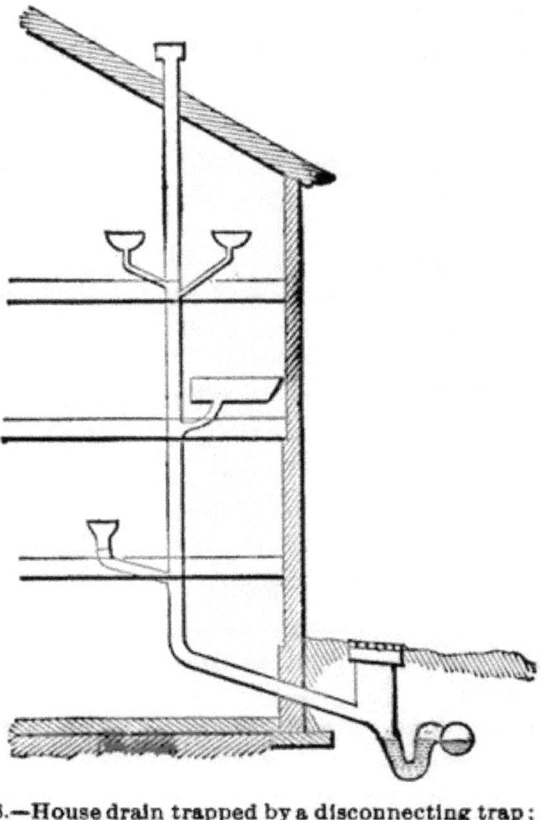

FIG. 38.—House drain trapped by a disconnecting trap; fixtures in the house left untrapped; soil pipe extended full size above roof.

the trap on the main drain and a fresh air pipe, or,

as it is sometimes called, a "disconnecting trap." The objection to this plan lies in the fact that soil and waste pipes of every house contain more or less foul air, which is not always expelled at the top of the soil pipe, but will enter the interior of the dwelling through untrapped fixtures. Even the short branches from fixtures become, in time, coated with a peculiar slime, emitting unhealthy gases; this is true in particular of the overflow pipes, which are insufficiently flushed and readily become the seat of fungoid growth. Noxious gases may, it is true, be withdrawn by connecting branch waste pipes to a hot flue. But the danger always remains that, at times, such flue ceases to draw; for instance, if the kitchen fire goes out over night, or, in the case of a steam coil placed in a flue to increase the draft, the steam may be shut off from Saturday afternoon to Monday morning. In such instance, what is to prevent the foul gases from entering through the fixtures into the house? Moreover, the practical difficulty is great of establishing a strong, *uniform* and constant downward draft through a multitude of untrapped plumbing fixtures.

Of the three methods of arranging the waste pipes of a dwelling, shown in Figs. 35, 36 and 38, the system illustrated in Fig. 35, showing a trap and fresh air inlet on the main drain, and a trap under every fixture in the house, is undoubtedly the *safest*, and therefore the best.

In the next chapter we will explain how accumulation of filth in traps, and therefore one of the chief objections against traps, may best be obviated by a judicious selection of a properly shaped trap. We shall also discuss under what conditions traps may fail, and how they can be made safe against back pressure, siphonage, evaporation of water, and absorption of gases by the water in the trap.

CHAPTER V.

DETAILS OF TRAPS.

IN a system of house drainage traps are necessary evils. I have explained in the last chapter why they must be used. They are evils because they tend to retard the flow of water through waste pipes, and, unless properly shaped, are apt to catch hair, lint, chips of straw or wood, and other articles, and retain more or less decomposing matter; for this reason, and where the water in the trap is not changed sufficiently often, they are the cause of annoying odors. Let us, therefore, inquire into the shape and character of traps used for house drainage purposes. To *all* traps the following cardinal principle, so well expressed by Mr. Hellyer, should apply: "No sanitary fitting, waste pipe, soil pipe or drain should be trapped in a way that will not admit of the whole of the water in such traps being entirely changed every time a good flush of water is sent into them." Although this rule applies to all kinds of traps, it is true above all of traps under urinals, slop-sinks and water-closets. These fixtures, therefore, should receive a liberal flush of pure water from a special cistern after each use. With kitchen and pantry sinks, laundry tubs, bath tubs and wash basins, the case is different. The usual custom is to empty these fixtures after use, without giving the waste pipe a subsequent flushing with clean water. The last water flowing from the fixture will remain, therefore, in the trap. Be this waste water from a bowl, a laundry tub, a bath or a sink, it is in any case *fouled* water which may emit noxious gases into the room, this depending, to a certain extent, upon the length of time during which the fixture remains unused. From this it is quite apparent that a judicious use of plumbing fixtures is all important in order to prevent traps becoming a serious evil.

When all washing is done, let the house-maid apply a thorough cleaning to all the tubs, and let her follow this with a few quarts of *clean* water from the faucet into each tub, and through the waste pipe into the trap. The same advice may be given with reference to the use of wash basins, bath tubs, etc. It is quite evident that *domestic cleanliness*, especially a proper care of fixtures, have much to do with the prevention of bad air in dwellings, but it would lead us too far to offer here more than these few pertinent remarks.

In considering the various traps in use it will be well to group them into the following classes: 1. Traps for house drains. 2. Water closet traps. 3. Traps for sinks, bowls and tubs.

Traps for fixtures as well as for drain pipes ought to be so shaped as to be *self cleansing*. A common pipe bent in the shape of an S, and therefore called S trap, of the same bore as the waste pipe, meets this requirement more thoroughly than any other kind. In the Minutes of Information on the "Drainage and Cleansing of Houses and Public and Private Edifices, etc.," published by the General Board of Health of England, in 1852, we find the following on traps for drain pipes :

"The best form of trap, the most simple, the least liable to derangement, and the most economical, and therefore the one to be recommended for house drains and for general adoption, is the Siphon water trap. For the ends of drains the siphon trap will be formed thus : (Fig. 39). These traps should, when practicable, be placed a little below the openings, so that the force of the fall of water may effectually discharge the previous contents."

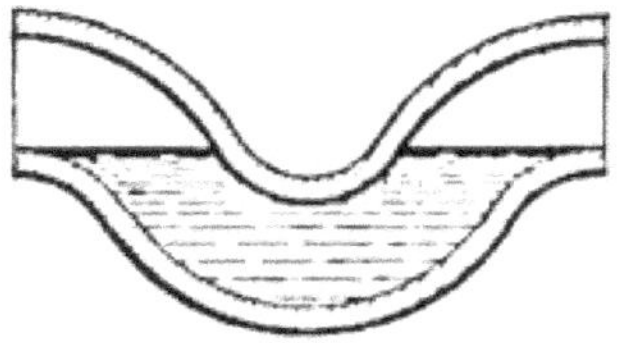

FIG. 39—Siphon water trap recommended by the General Board of Health, of England.

Traps for House Drains.

The earliest traps placed on house drains to separate the house from cesspools or sewers were probably flap valves, such as shown in Fig. 40, but

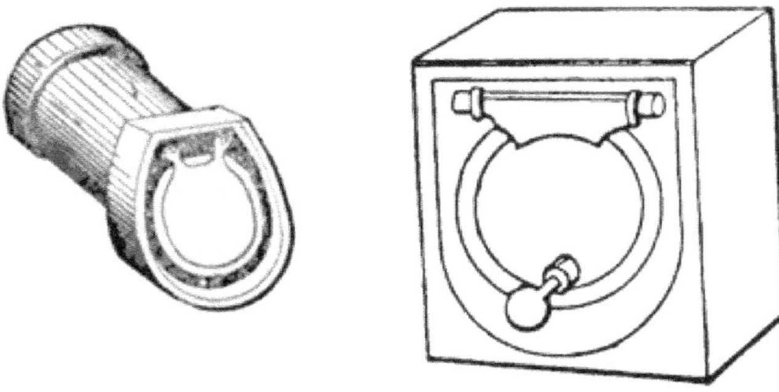

Fig. 40—Flap Valve for House Drainage.

it was soon recognized that even light flaps would tend to detain coarse waste matters and cause obstructions in the house drain as shown in the sketch. (Fig. 41.) Moreover, none of the flap

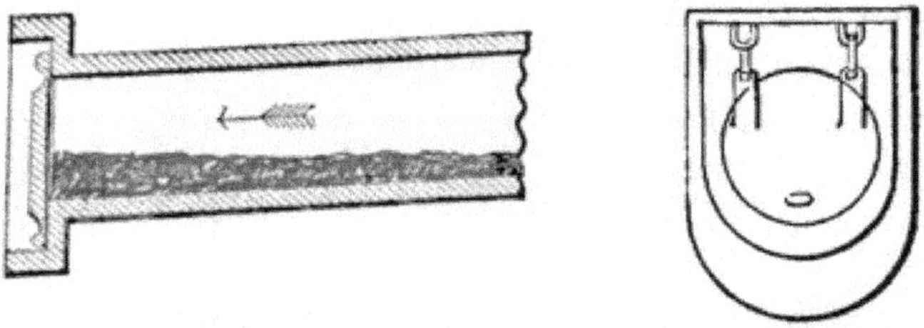

FIG. 41.—Flap Valve for house drain (taken from General Board of Health Report, 1852).

valves on drains would form an air-tight seal against gases of decomposition.

An equally objectionable form of trap is the "cesspool trap," or "mason's trap" (Fig. 42), so

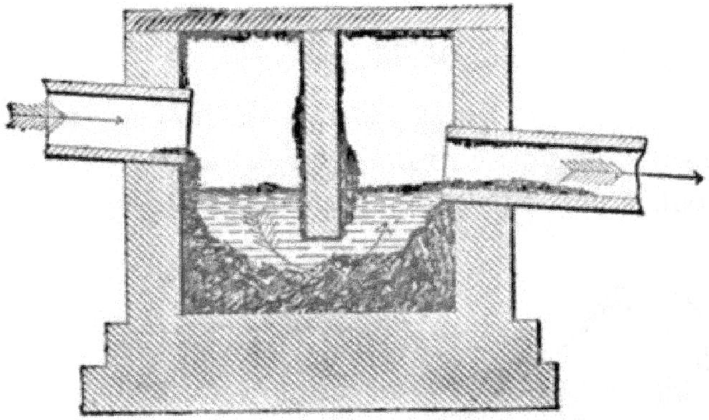

FIG. 42—Cesspool or Mason's Trap.

commonly found in old city residences and country mansions, and invariably filled—often choked—with the worst kind of putrescent matter. Most traps now used for house drains have the siphon-shape (Fig. 39) and are generally provided with an inlet for fresh air on the house side of the water seal. Such traps are made in cast-iron and in earthenware, and are placed near the front wall in the cellar, or outside of the house, in which case proper precautions should be taken to protect the trap from freezing and to make it accessible for inspection and for cleaning purposes. Earthen traps should be highly glazed to present a smooth surface, while the iron traps may be coated with the black or the white porcelain enamel.

In most traps for house drains the fresh air inlet on the house side of the water seal is combined with the trap in one piece. Sometimes this inlet is enlarged to an air-chamber, and the trap is then generally called a "disconnecting trap." Most of these, as we shall see, are of English make, and used there extensively, while none but the simpler traps are used in the United States.

There is a radical difference between the English house drainage system and the system used with us, which may be readily explained by the difference of climate. It is a cardinal principle with English sanitary engineers to locate soil pipes outside of the house, and further, to separate water closet wastes from most other wastes of the household. Waste-pipes from lavatories, bath tubs, sinks, etc., are required to have no direct connection with a foul water drain; they must discharge over open gullies, which are trapped and connect to the house drain. The severity of our climate would prohibit such an arrangement in all but the Southern States. We *must* keep soil and waste pipes inside of a dwelling, and, on the other hand, do not for a moment hesitate to connect bath or bowl wastes to a soil pipe, provided the latter is efficiently ventilated and the fixtures safely trapped. I have made mention of the English practice because many of the drain traps illustrated are of English make, and thus their arrangement will be more readily understood.

I will now briefly describe and illustrate some of the numerous drain traps used in modern works of house drainage :

Fig. 43 and Fig. 44 are running traps of cast-iron, manufactured for use with plumber's soil

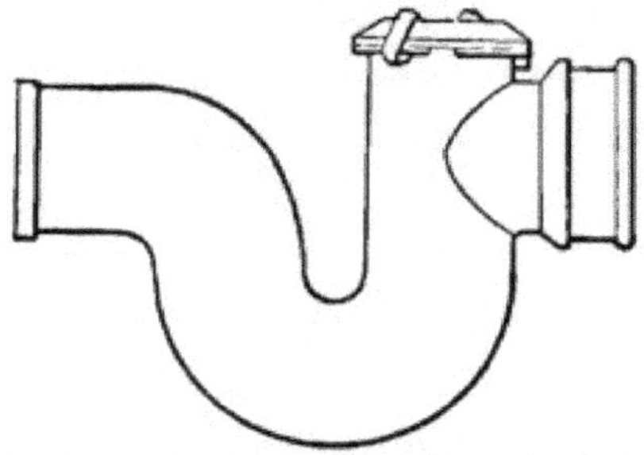

FIG. 43—Mott's Running Trap, with cleaning hole and cover.

pipe by the J. L. Mott Iron Works, and others. Fig. 43 illustrates a trap with a hand-hole for cleaning purposes, closed air-tight by an iron

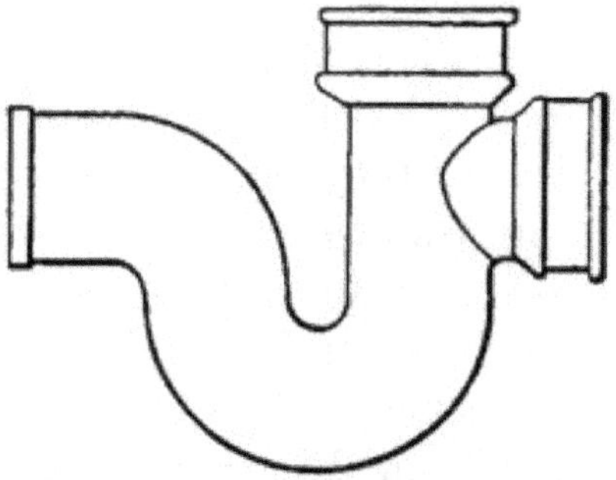

FIG. 44—Mott's Running Trap, with opening for fresh air.

cover, set in Portland cement. Fig. 44 shows a trap with opening for a fresh air pipe on the house side of trap.

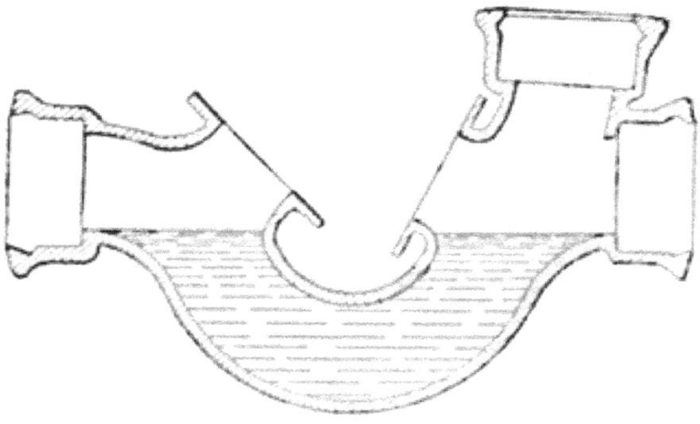

FIG. 45—Durham Trap, with cleaning holes and opening for fresh air.

Fig. 45 illustrates a running trap for house drains, manufactured by the Durham House Drainage Company. It is provided with two large hand-holes for cleaning purposes, closed by iron plates bolted to the flange, the joint being made tight with red lead and putty. These plates are supposed to be removed in the illustration. The trap has also a large opening on the house side for

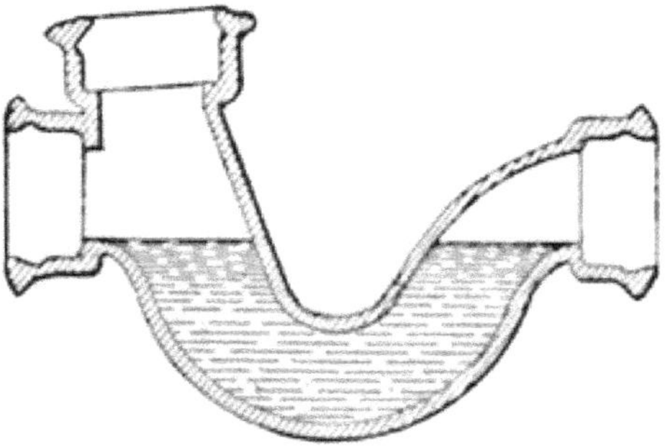

FIG. 46—Durham Trap, with fresh air inlet only.

a fresh air pipe. Fig. 46 shows the same kind of trap with fresh air inlet, but without cleaning hand-holes.

Fig. 47 is a representation of the Stewart trap, made in earthenware. In addition to the opening B, intended for cleaning purposes, or else to introduce fresh air into the house drain, it has a second opening, D, to which a vent pipe may be attached, leading to the open air, remote from windows, or else extended up to the roof. In case of pressure from the sewer, which may occur at times of sudden rain-falls, or with sewers, exposed to the influence of the tide, and for other reasons, the first seal, which is not quite accurately shown in the cut, may be forced, and sewer air would then escape through the opening D instead of into the house pipes.

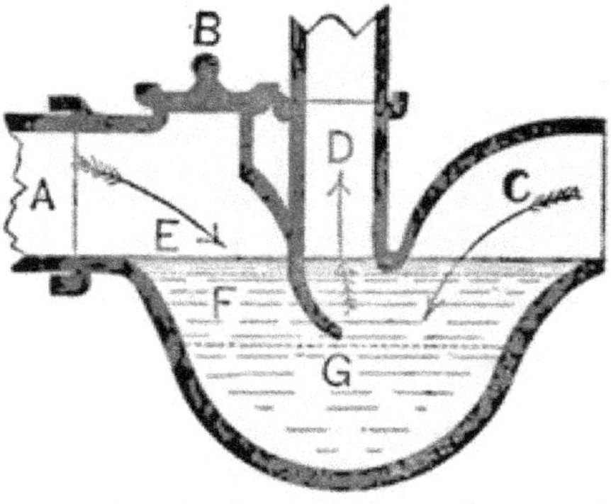

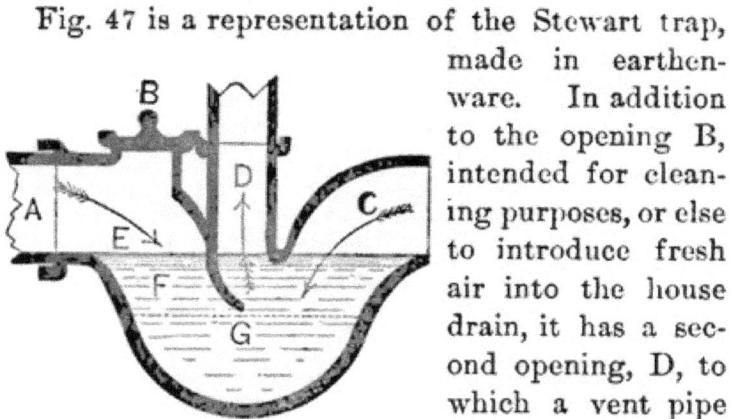

FIG. 47.—Stewart's earthenware House Drain Trap.

Fig. 48 shows a double trap on the line of the main drain, which, however, should not be used except where there is apt to be excessive pressure from the sewer. If it is used, a vent pipe should

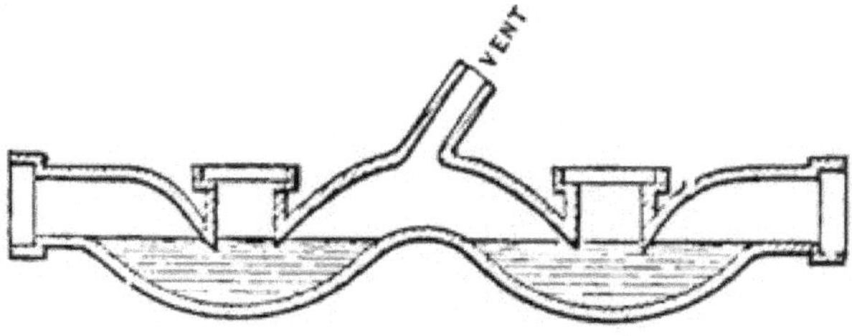

FIG. 48.—Double trapped House Drain, with vent pipe between first and second trap.

be placed between the first and second trap, leading up to the roof.

Fig. 49 illustrates a drain trap, located outside

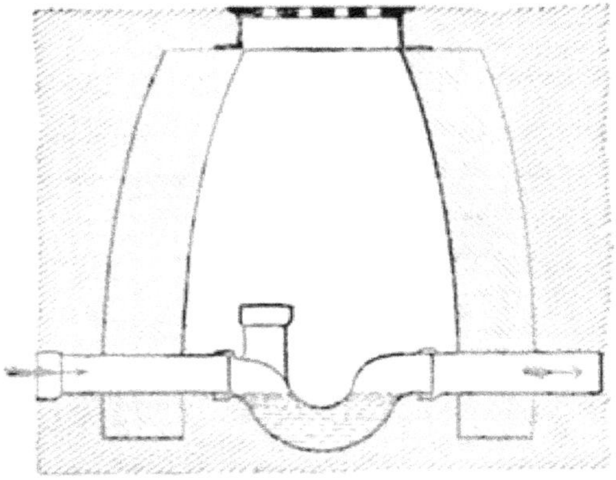

FIG. 49.—Trap and opening for fresh air located in a manhole.

of the house in a man-hole, and having an opening serving as a fresh air inlet. The top of man-hole should, then, be covered with a perforated cover or grating.

Fig. 50 illustrates in section an unsatisfactory

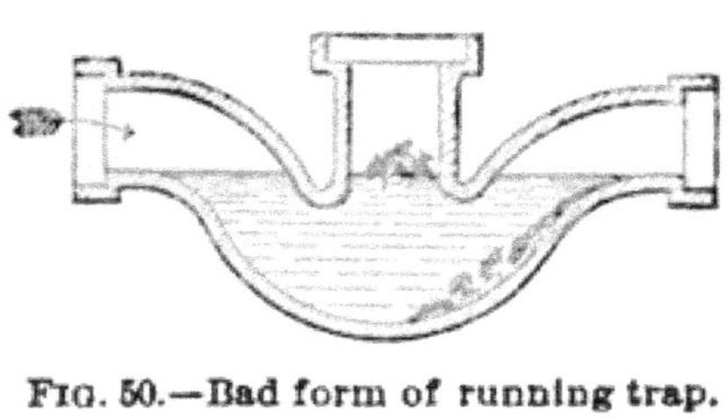

FIG. 50.—Bad form of running trap.

shape of trap, commonly made in earthenware, and indicates the manner in which this trap soon accumulates filth.

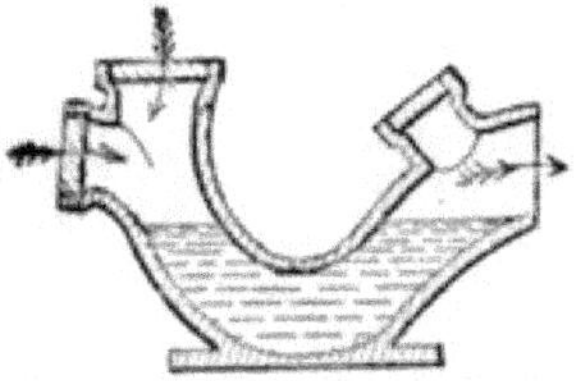

FIG. 51.—Suitable Stoneware Trap for house drains.

Fig. 51 is a sketch of a suitable stoneware trap for house drains, recommended in the Model By-Laws of the Local Government Board.

The same by-laws recommend the disconnecting trap, Fig. 52, for house

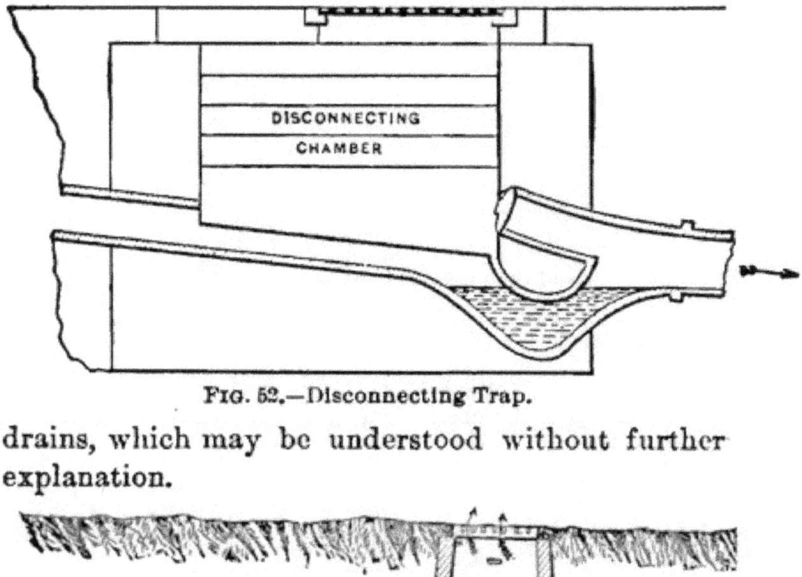

FIG. 52.—Disconnecting Trap.

drains, which may be understood without further explanation.

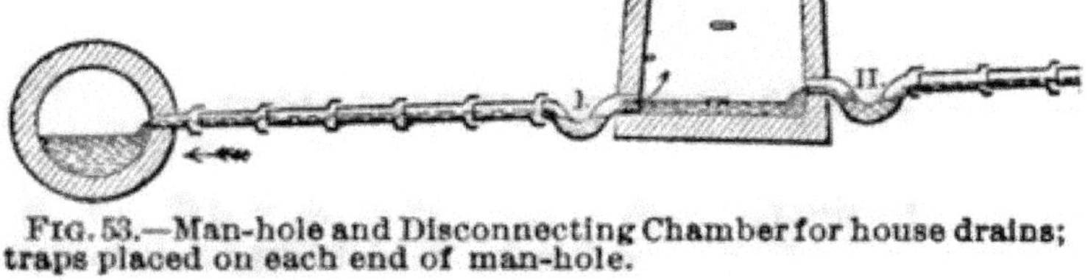

FIG. 53.—Man-hole and Disconnecting Chamber for house drains; traps placed on each end of man-hole.

A house drain may be disconnected from the

sewer or cesspool in the manner indicated in Fig. 53. Should gases force their way through the first trap, they would escape at the man-hole, and the second trap effectively prevents entrance of sewer air. Such an arrangement is feasible only for suburban or country dwellings; it is not much in use.

A trap, with large-sized fresh air inlet, and called a disconnecting trap, is shown in Fig. 54. This

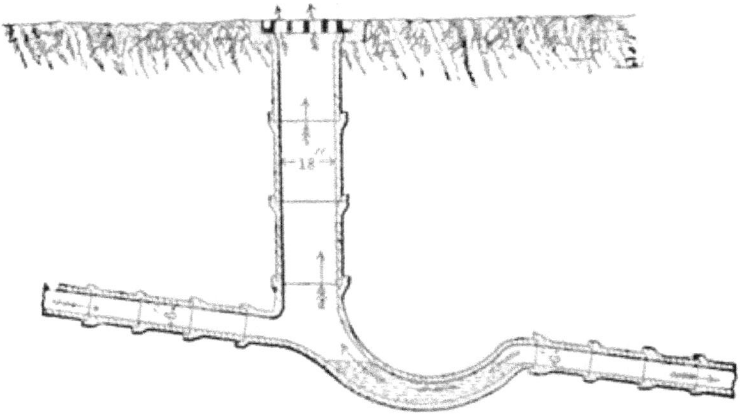

FIG. 54—Disconnecting Trap.

arrangement was first recommended by Dr. Buchanan, of England, and afterwards by the Massa chusetts State Board of Health.

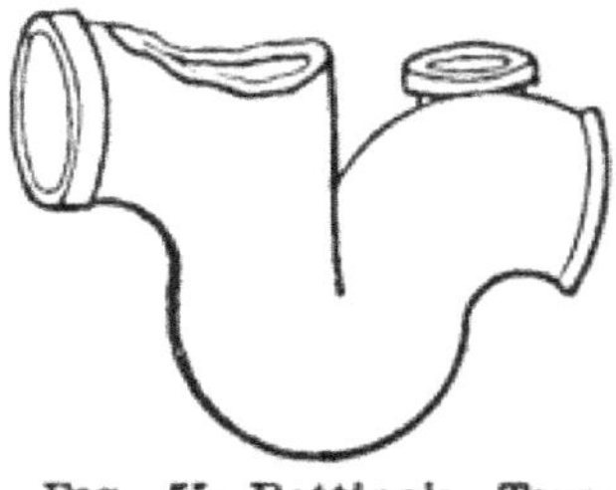

FIG. 55—Botting's Trap, with fresh air inlet.

Fig. 55 illustrates a recently designed trap, manufactured in cast-iron, with oval fresh air opening, known as Botting's trap.

Figs. 56 and 57 are representations of the Bavin and the Redhill Traps; both are made in stoneware. The shape of the Redhill Trap is apparently a very good one.

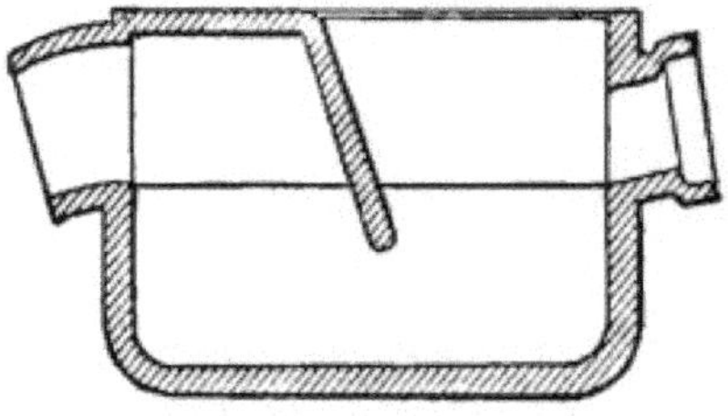

FIG. 56.—Bavin Trap.

FIG. 57.—Redhill Trap.

Fig. 58 is the "Cascade" Action Trap, patented to Mr. Buchan, of Glasgow. It is made in stoneware, and provided with a cleaning hole, and with an inlet for fresh air. There is a vertical drop of about two inches from the house drain side and into the water of this trap, and such a drop is generally regarded as an advantage, as it tends to "break up and carry away the fæces" more readily. In a recent letter to the author, Robert Rawlinson, Esq., wrote:

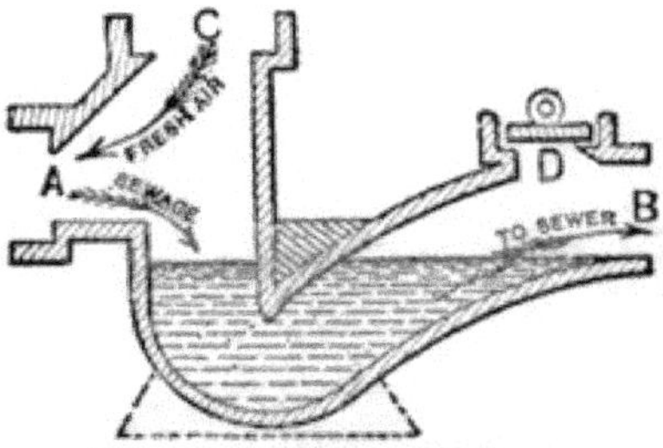

FIG. 58.—Buchan's Trap.

"With respect to traps there are various forms dependent on water. The inverted siphon is one of the most common. In this form of trap it is an advantage to have the outlet lower than the inlet, and not on a level. Where the difference in level will allow, a vertical fall from a house drain is useful into the small hand-chambers over the head of the inverted siphon, and in America this should be placed well down so as to be out of the reach of frost."

Moleswerth's Trap, Fig. 59, is one of the earlier

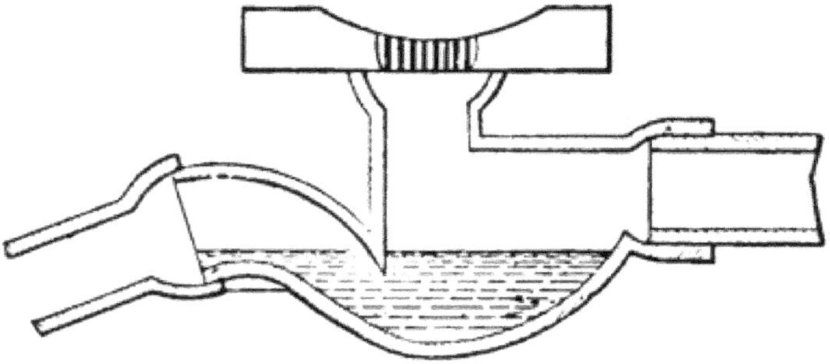

Fig. 59—Molesworth's Trap.

forms of disconnecting traps, which need little explanation. The water level is not correctly shown in the drawing of this trap.

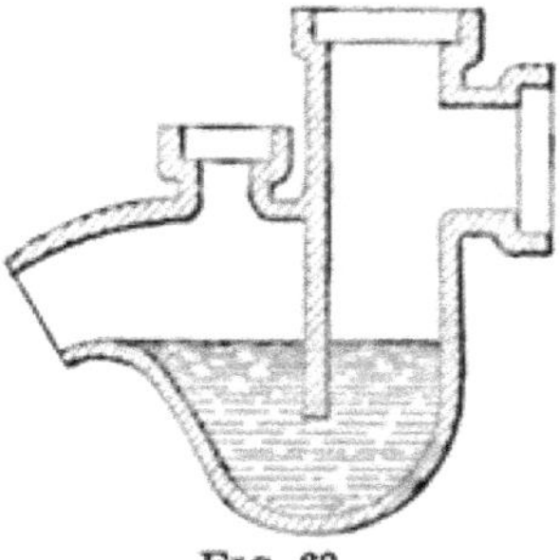

Fig. 60.

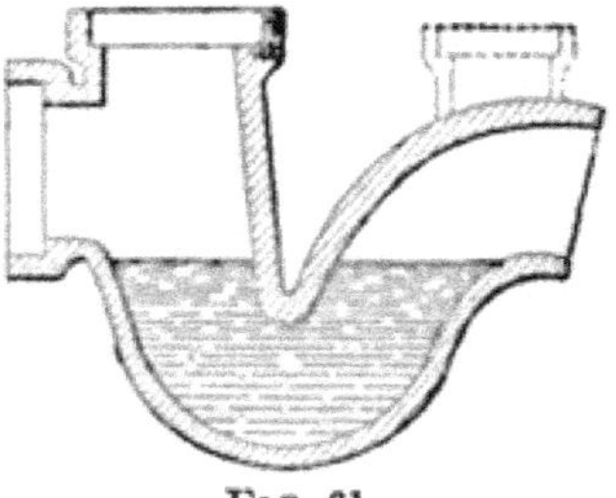

Fig. 61.

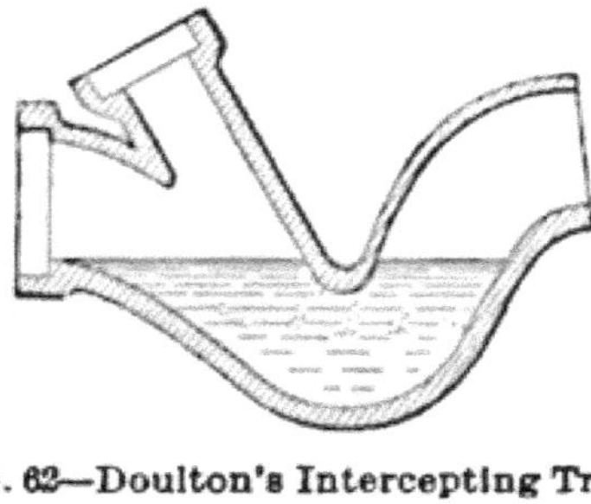

Fig. 62—Doulton's Intercepting Traps.

Figs. 60, 61 and 62 are illustrations of three different intercepting traps, manufactured in stoneware by Doulton, of Lambeth, England.

Figs. 63 and 64 show two forms of the "Eureka" Sewer Air Trap, the suggestion of the well-known

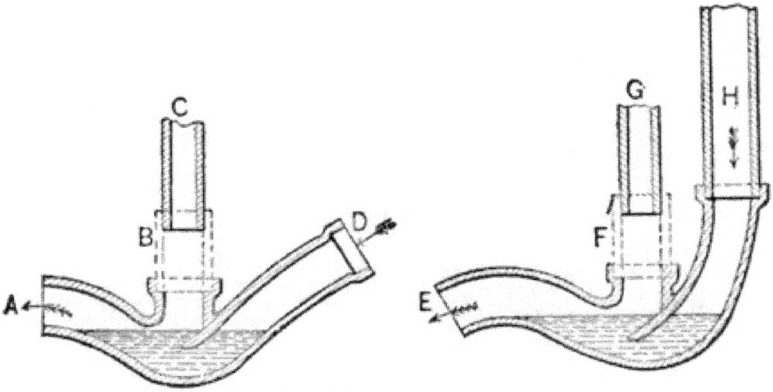

FIGS. 63 and 64—Eureka Traps.

sanitarian, Dr. P. Hinckes Bird, manufactured in stoneware by James Stiff & Sons, Lambeth. The curve of the siphon is an easy one, and the curved dip will facilitate the scouring and flushing of this trap. The opening beyond the dip may be used as a cleaning hole, or else it may be connected with a vent pipe carried up to the roof, and intended to remove gases from the public sewer or cesspool.

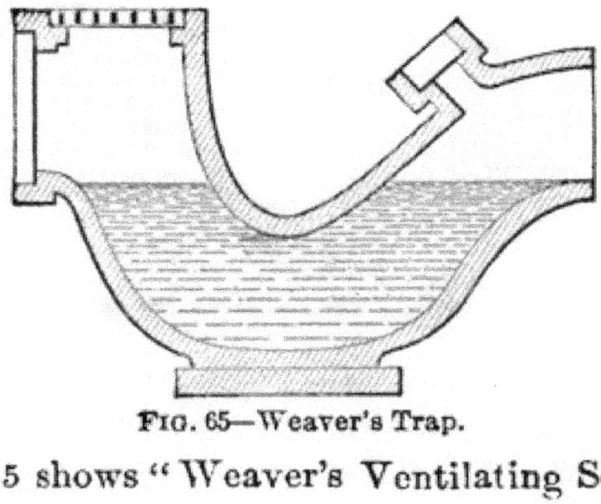

FIG. 65—Weaver's Trap.

Fig. 65 shows "Weaver's Ventilating Sewer Air

Trap, manufactured by James Stiff & Sons, of Lambeth, England. This is a most convenient

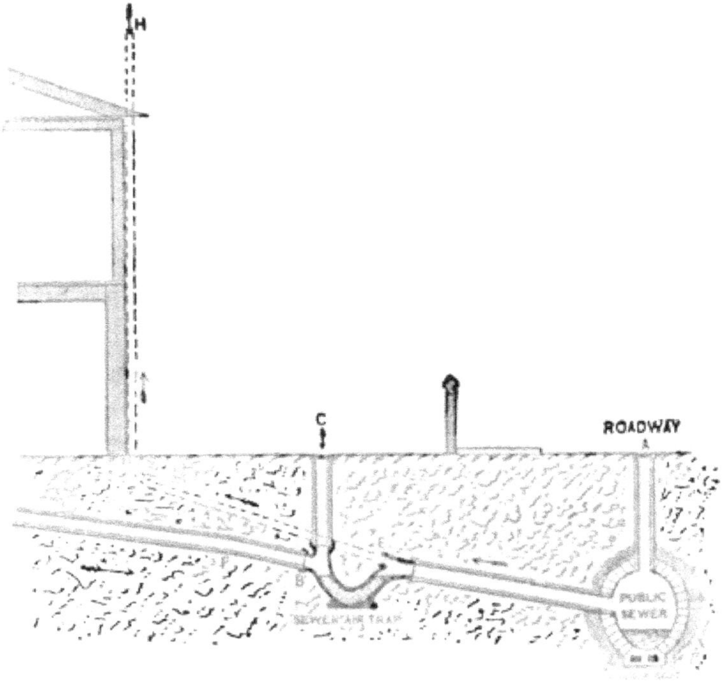

FIG. 65a.—Ventilation of the house drain and the sewer, as effected by Weaver's Trap.

form of a disconnecting trap. It has a fresh air inlet covered with a perforated grating, and on the sewer side of the dip, a junction, which should be connected to a pipe leading up to the roof on the *outside* of the house, to ventilate the sewer. Where this trap is placed much below the surface, to bring it out of reach of the frost, the fresh air inlet should be brought up to the surface by pipes, such as shown at C in Fig. 65a.

Fig. 66 is another trap made by Stiff & Sons, of Lambeth, and called the registered "Intercepter" Sewer Air Trap. This trap is made in stoneware, and has at its upper part three openings. The trap has a double water-seal, the nearest to the sewer being only 2 inches deep, while the second one is 7 inches in depth. The first dip effectually disconnects the house drain from the air of the sewer or cesspool. Should there be an excess of pressure from either, the first water seal may be forced, but the foul gases will then find an exit at the middle or foul air outlet, which should be carried up to the roof of the building by a pipe on the outside of the house. No foul gases from the sewer will, under such conditions, pass through the second stronger water seal. *Fresh air* will continually enter at the inlet nearest to the house, thus establishing the desired constant change of air in the soil and waste pipes. The opening nearest to the sewer is intended for inspection and cleansing purposes. Where the trap is located much below the surface, an earthern pipe should be carried from the openings to the surface, the first one being covered by an open grating, the second one by a tight cover.

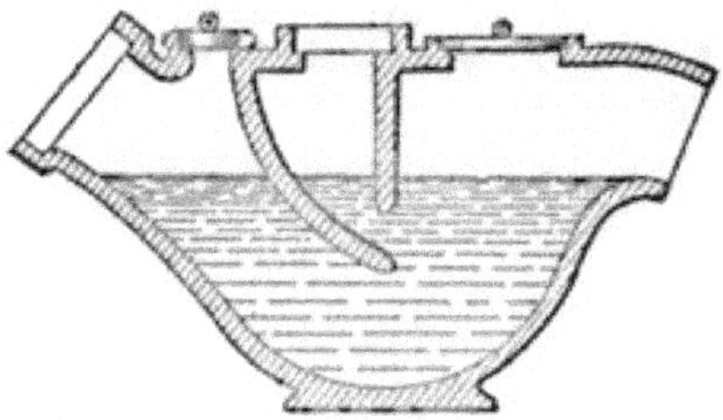

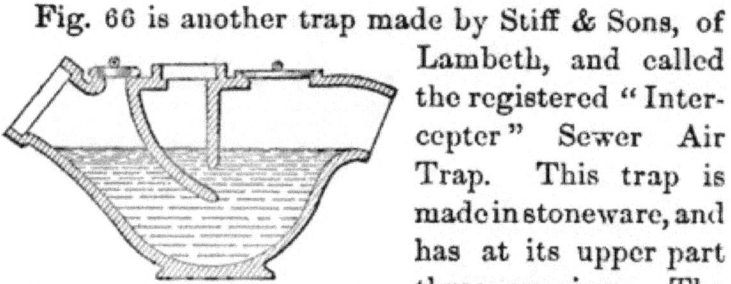

FIG. 66.—Stiff's "Intercepter" Sewer Air Trap.

Jennings' ventilating cesspool or drain trap, Fig. 67, is very similar in shape and identical in design to Buchan's trap. This trap has the above-mentioned drop of several inches from the house drain to the level of the water in the trap. It is

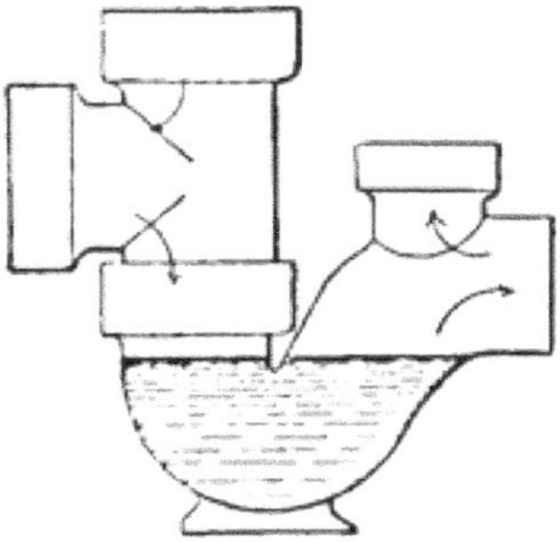

FIG. 67—Jennings' Ventilating Drain Trap.

designed by the well-known manufacturer of sanitary fittings, George Jennings, of London, and is made in strong, highly-glazed, vitrified stoneware. As the patentee says, "it is designed for introduction, *not at the foot of a soil pipe, or in a line of drain*, but at the point of junction with the cesspool or sewer, the proper ventilation of which can be provided in connection with the smaller socket. At the large socket a grating may be used, or a ventilating pipe inserted, and carried to a convenient height to allow the escape of air driven down by descending waters, and to admit fresh air at the lowest point in the line of drain. The trap being formed in two pieces, the socket may be turned round in any required angle to the line of drain, and several pipes may be connected to one outfall by the substitution of a junction piece having two or three inlets."

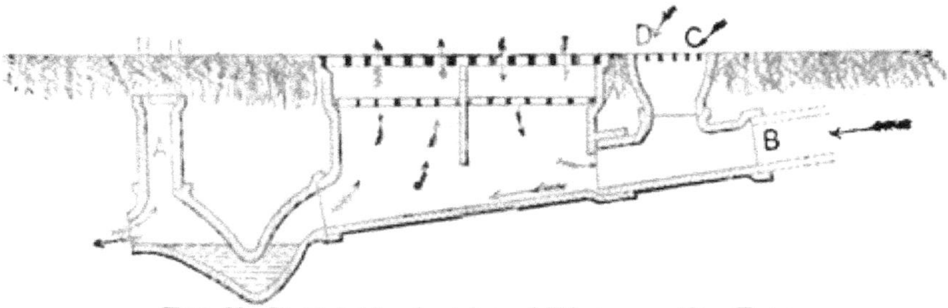

FIG. 68—Potts' Air-chambered Disconnecting Trap.

Fig. 68 gives a sectional view of Potts' Edinburgh air-chamber sewer trap, manufactured by Potts & Co., of Handsworth, near Birmingham, England. It is an efficient disconnecting trap, well adapted for mild climates, and takes its denomination from the large air-chamber, which is covered by a double grating. On the lower grating may be placed charcoal or other disinfectants, which, however, tends to interfere with a proper air-current. The air-chamber has a dividing diaphragm, intended to assist in creating a current of air through the trap. The house drain discharges at the head of the chamber, at B. Leader, sink and gully wastes discharge into side openings in the air-chamber, or else over the grating at C D. Beyond the water seal siphon-trap is an opening, A, which can be used for cleaning purposes, or else may connect to a pipe leading to the roof, and helping to ventilate the public sewer. Where this trap is placed much below the surface, the air-chamber may be extended upwards to the surface.

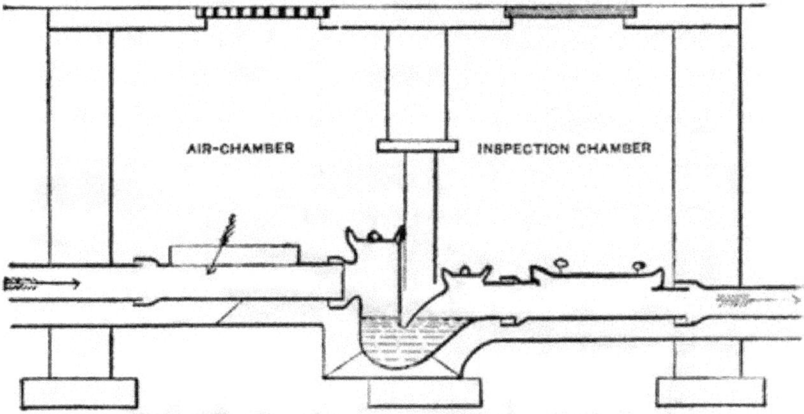

FIG. 69—Buchan's Disconnecting Trap and Chamber.

Fig. 69 shows a vertical section of Buchan's

disconnecting trap, consisting of two man-holes, with a Buchan "Cascade action" trap between both on the line of the drain. The man-hole next to the sewer serves for cleaning and inspection purposes; the man-hole next to the house serves as an air-chamber, and the drain pipe has a large opening through which fresh air enters the house drain. Buchan has also constructed a disconnecting trap placed in a single man-hole.

Fig. 70 illustrates the Croydon Siphon Trap, which is also a disconnecting trap, but its shape is such as to make it hold a large quantity of foul water, which is not readily expelled. It is consequently not a self-cleansing drain trap.

FIG. 70—Croydon Siphon Trap.

Fig. 71 shows the ventilating drain siphon and sewer intercepter, patented by Mr. Hellyer, and manufactured in stoneware. "These traps," Mr. Hellyer says, "are specially constructed for intercepting, or rather *disconnecting* sewers and sewage tanks from the house drain. The trap consists of a *round pipe*, shaped in the form of the letter V, giving it a *water seal* or *dip* of about 3 inches, and the body of the trap is comparatively of smaller diameter, to prevent any filth collecting in the trap, and also to allow the water in the trap to be more easily driven out by the flushes from

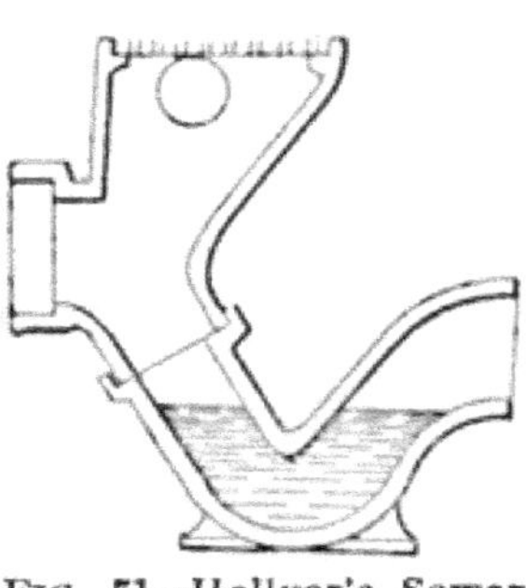

FIG. 71—Hellyer's Sewer Intercepter Trap.

the drain." There is also a fall of about 6 inches from the drain to the level of the water in the trap. The upper part of the trap is enlarged for the admission of fresh air, and carried up to the surface, where it should be covered with an open grating.

Mr. Hellyer has devised another trap, which he calls the "Soil pipe Disconnector," shown in Fig. 72, and also a trap which he calls the "Combina-

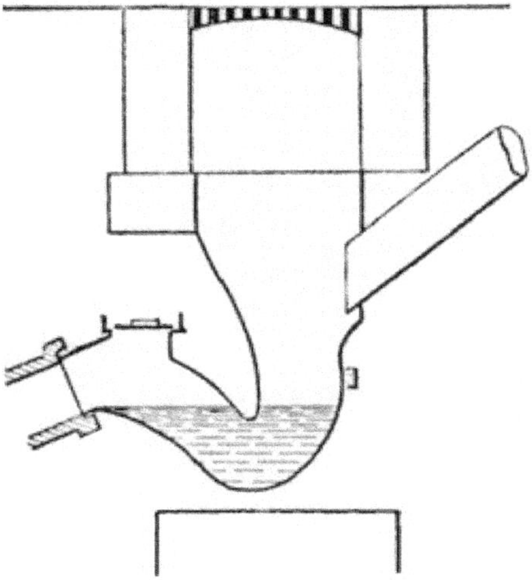

FIG. 72.—Hellyer's Soil-pipe Disconnector.

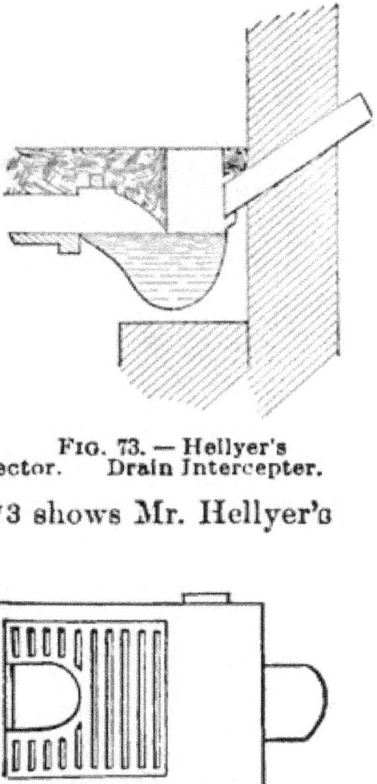

FIG. 73. — Hellyer's Drain Intercepter.

tion Soil Pipe Trap." Fig. 73 shows Mr. Hellyer's Drain Intercepter Trap.

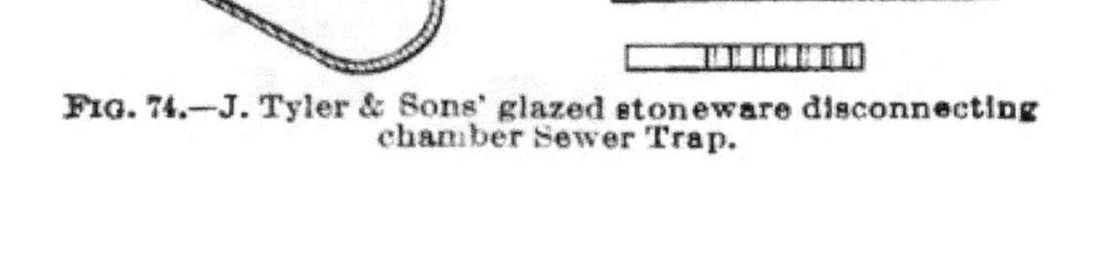

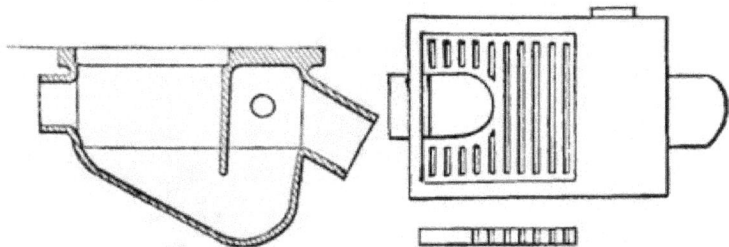

FIG. 74.—J. Tyler & Sons' glazed stoneware disconnecting chamber Sewer Trap.

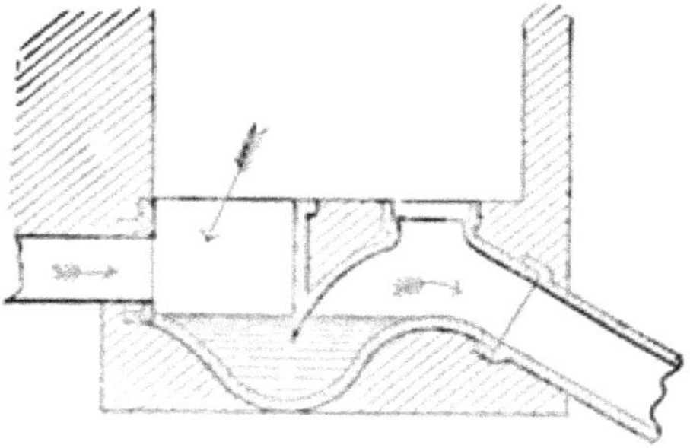

FIG. 75.—J. Tylor & Sons' disconnecting chamber Sewer Trap.

Two disconnecting chamber sewer traps, made in glazed stoneware by J. Tylor & Sons, are illustrated in Fig. 74 and Fig. 75, which hardly need any detailed description.

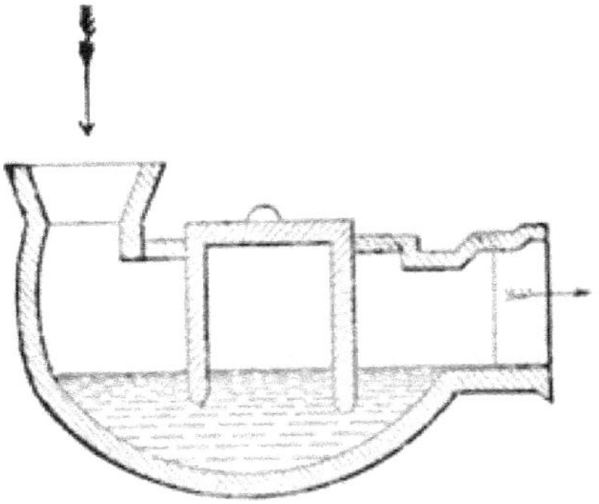

FIG. 76.—Cottam's Trap.

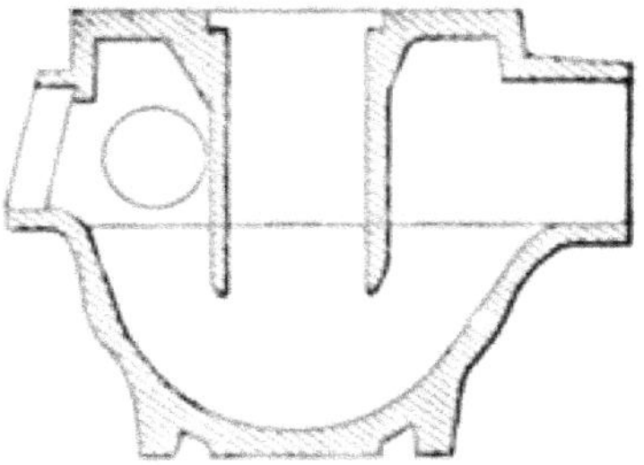

FIG. 77.—McLandsborough's Trap.

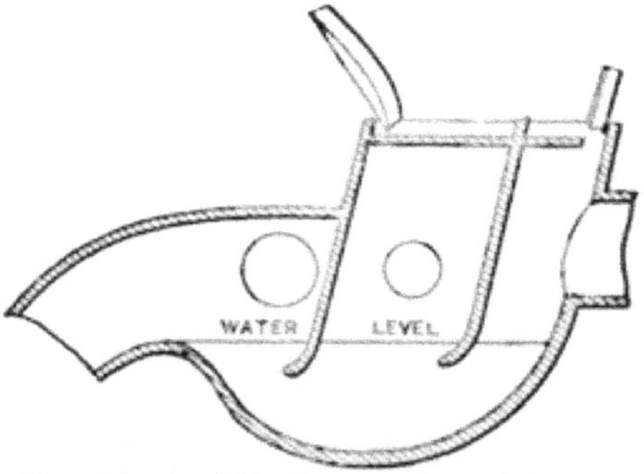

FIG. 78.—Dodd's Patent Stench Trap.

Cottam's trap, Fig. 76, McLandsborough's trap, Fig 77, and Dodd's Patent Duplex Stench trap, Fig. 78, have each a double dip or water seal, but I should hardly call such traps self-cleansing, as grease and solids will, after some use, accumulate on the surface in the central chamber between the two diaphragms, especially in the two first-named traps. I can see no advantage of any of these traps over some simpler traps, described above.

Fig. 79 is a vertical section of Stidder's inter-

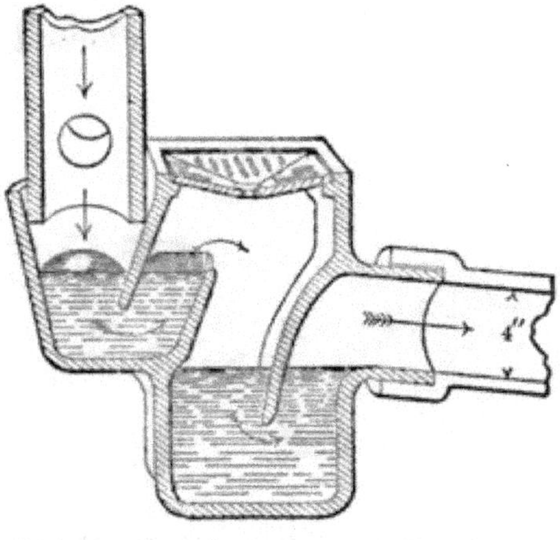

FIG. 79—Stidder's Disconnecting Trap.

cepting and disconnecting trap. It has a double water seal, an air-chamber, and a surface grating between both. It may answer for surface water and wastes from hydrants in yards, but I do not think it would be self-cleansing, when used for household waste water, even excluding, as the patentee does, water-closet wastes.

The same objection may be made against Hellyer's "triple dip trap or drain intercepter" (Fig. 80), made in stoneware. There are *three* water-dips in this trap, and thus the security from gases from

FIG. 80.—Hellyer's Triple Dip Trap.

the sewer or cesspool is largely increased, but, it must be conceded at the expense of simplicity in construction.

The firm of J. G. Stidder & Co. (London Sanitary Engineering Works) manufacture a large variety of intercepting traps for house drains, but it would lead us much beyond the scope of this little volume to illustrate all of them. We refer to the handsome illustrated catalogue, issued by the firm.

Fig. 81 represents Copley Woodhead's double

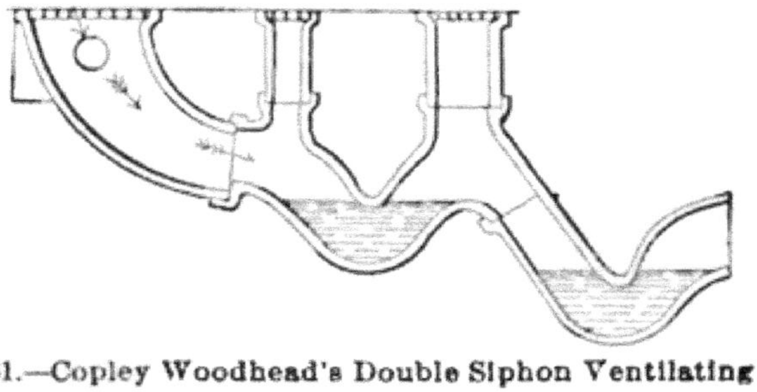

FIG. 81.—Copley Woodhead's Double Siphon Ventilating Sewer Trap.

siphon ventilating sewer trap. Should the trap nearest to the sewer be forced, the gases find a ready exit at the first air shaft, the second siphon effectually excluding them from the house. The second air shaft serves to admit air to the house drainage system.

Fig. 82 illustrates two forms of soil pipe traps,

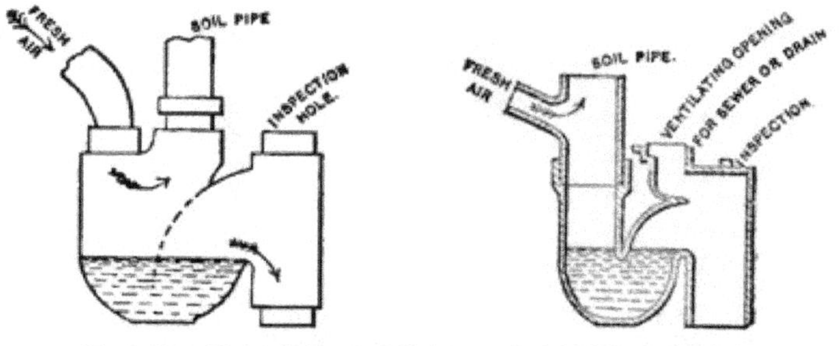

FIG. 82.—Two forms of Banner's Soil Pipe Traps.

used by Mr. Banner in his patented system of soil pipe ventilation. The practice of trapping soil pipes at their foot, and introducing at the same point fresh air from the outside, is restricted to a few systems of soil pipe ventilation, which have, thus far, found no favor in the United States.

A few illustrations of gully traps and sink traps may close our list of traps for house drains. Fig. 83 shows the "Weatherly disconnector" waste water trap, used for rain leaders, surface water, and for waste pipes of sinks, bath, lavatories, but not for soil pipes. It may be of service in England, where water-closet wastes are kept separate from wastes of sinks, bath and lavatories, and where the latter are required to discharge over an open grating. Such traps are

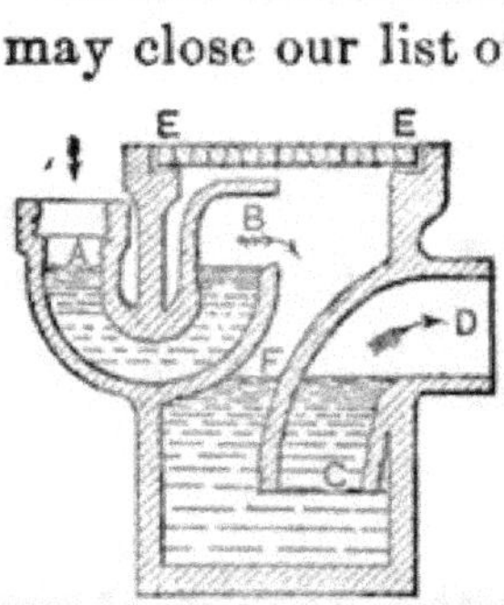

FIG. 83.—Weatherly Sink Trap.

not adapted to American methods of house sewerage. A well-known trap of this kind is Mansergh's trap.

Figs. 84 and 85 illustrate Lovegroove's patent

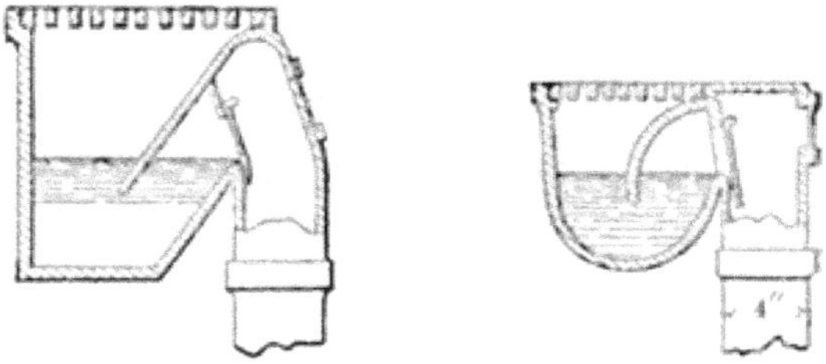

FIGS. 84 and 85.—Lovegroove's Patent Drain Traps.

drain traps, the former to be used for stable or yard drainage, the latter for areas. Of these traps, which are really mechanical traps, having a flap valve, the manufacturer says: "These traps are, under all circumstances, equally efficient with or without water, the absorption of sewer gases being prevented by the valve, which also effectually prevents the escape of impure air from the drains, in the event of the loss of water seal by evaporation or other cause."

Bellmann's patent gully is shown in Fig. 86, its chief advantage being the P or S trap in place of the bell trap ordinarily used. The top piece can be turned any way to suit all localities. It has side openings to receive sink or bath wastes as shown.

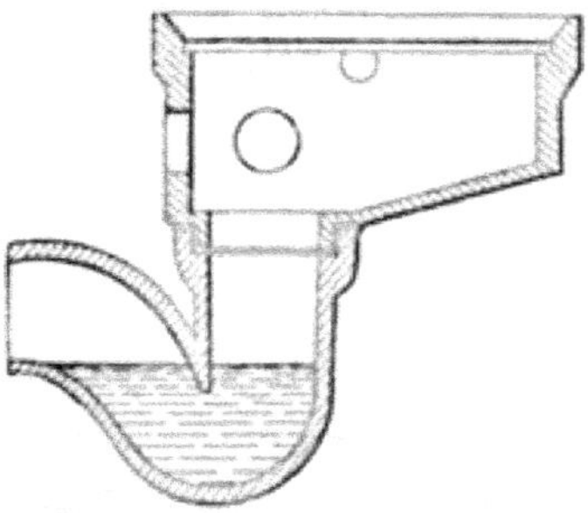

FIG. 86.—Bellmann's Patent Gully.

Lastly, we mention Jennings' improved tidal valve

trap for the outfalls of drains or sewers, subjected to the influence of high tides, heavy rains, etc. It is a mechanical valve, and Fig. 87 "is a sectional

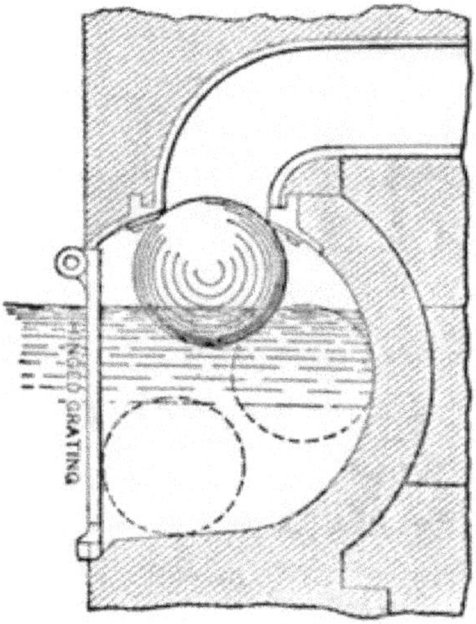

FIG. 87.—Jennings' Improved Tidal Valve.

view of it, showing its position when the stream or tide rises up to or above the point of discharge, the buoyancy of the ball causing it to be carried into the orifice of the discharging pipe, and against an india-rubber valve seat, forming a perfectly tight joint, greater pressure from below tending only to increase its security. A grating, secured and hinged, admits of easy access to the valve or chamber at any time when required. Under ordinary conditions, the ball, which is of copper, floating or resting in a chamber formed in cement or masonry of larger dimensions than the discharging pipe, offers no obstruction to the free escape of waste or storm waters, while the 'drop' from the pipe effectually prevents the lodgment of anything to obstruct the proper closing of the valve."

Such tidal valve is useful for the protection of property below the level of high tides, and also where there is, at times, a back pressure from the sewer, in case of heavy rain-falls. It must not be forgotten, though, that in using such valve, the lower or outlet part of the house drain must be increased to a capacity equal to the amount of sewage discharged from the house during such period of high tide, otherwise a backing up of sewage into the cellar and through basement fixtures may occur.

Traps for Water Closets.

The trap most commonly used for water closets until a few years ago was the well-known D trap. It presents inviting recesses for the accumulation of grease and filth (see Fig. 88), and should not be tolerated under any circumstances whatever in a house which makes any pretense to be in a sanitary condition. The fact of its having a large cleaning screw (Fig. 89) does not make it any more acceptable, for such a screw is inconveniently located, below the floor and out of sight, and is consequently never thought of. Fig. 90a illustrates a vertical section, and Fig. 90b a plan of Tylor's galvanized iron closet D-trap, in which the cleaning screw is located on top, so as to be readily removed.

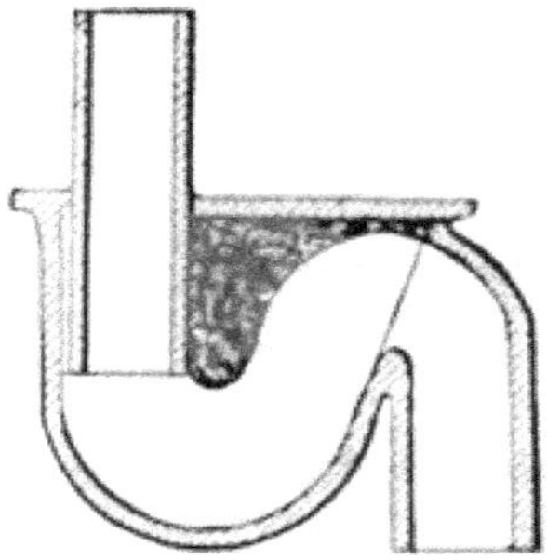

Fig. 88.—D-Trap, with filth accumulation.

Even this trap is objectionable, as it is not self-cleansing.

FIG. 89.—D-Trap, with brass cleaning screw on the side.

The "Helmet" trap, Smeaton's "Eclipse" trap (Fig. 91), and the Adee Patent Stench Trap (Fig. 95), are not much better designed with respect to

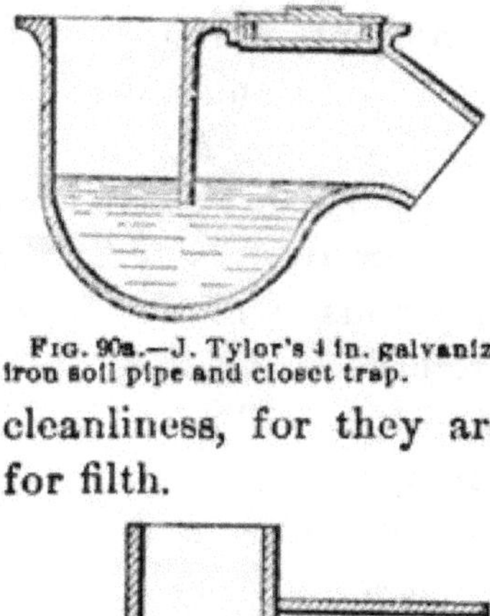

FIG. 90a.—J. Tylor's 4 in. galvanized iron soil pipe and closet trap.

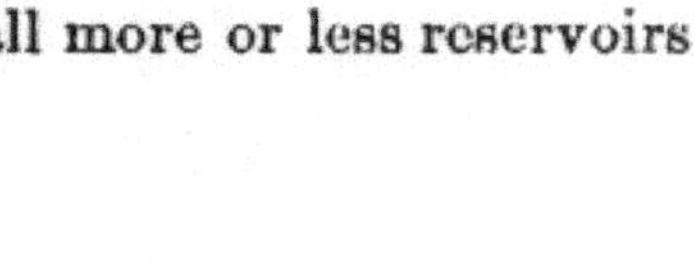

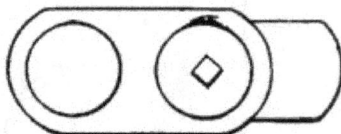

FIG. 90b.—Plan of same.

cleanliness, for they are all more or less reservoirs for filth.

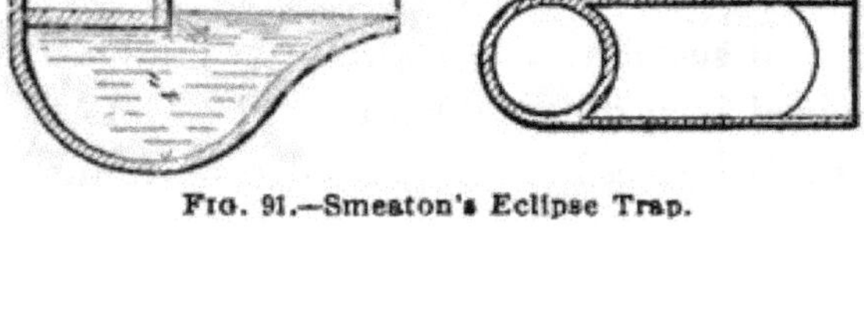

FIG. 91.—Smeaton's Eclipse Trap.

Fig. 92 illustrates, in view and in section, Hellyer's Mansion Trap for water closets, and Fig. 93 a, b, c, Hellyer's Anti-D-Trap, for which latter trap the inventor claims that it is not easily siphoned. There are two sizes of the Anti-D-Trap, both of them, but especially the smaller one (Fig. 93b), designed to hold the least quantity of water consistent with a sufficient water-seal. Mr. Buchan of Glasgow proposes the Anti-D-Trap (Fig. 94), which he believes is safer from siphonage than the Anti-D-Trap of Mr. S. Hellyer, as it has a large air chamber on the sewer side of the trap.

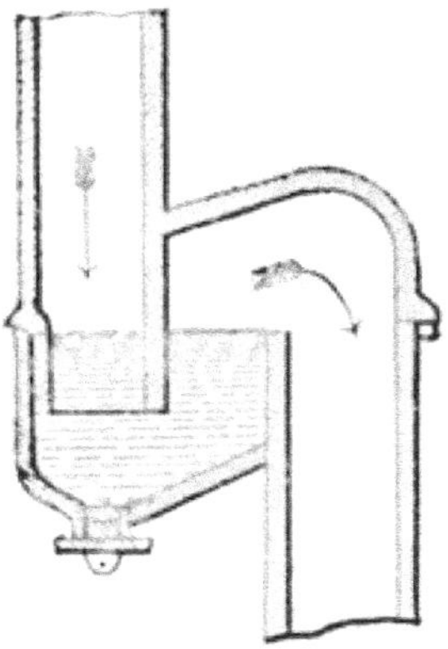

FIG. 95.—Adee's Patent Stench Trap.

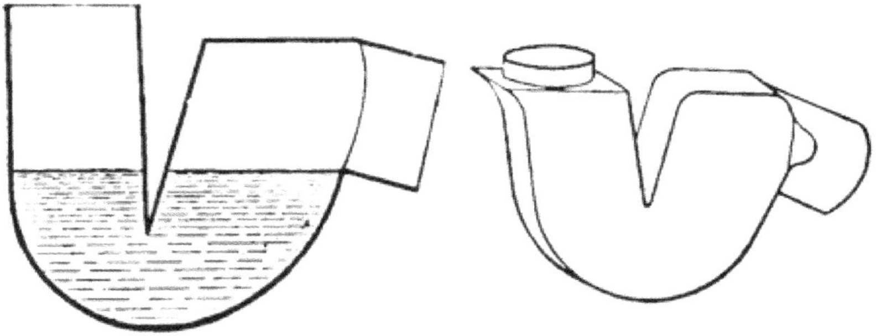

FIG. 92.—Hellyer's Mansion Trap.

Water closet traps should not have too large a dip or seal, for otherwise it is difficult to drive paper and solids out into the soil pipe. The less quantity of water such a trap holds, with the same depth of water seal, the better will it be, for it will then be possible to change its contents entirely at each flush.

No mechanical trap has as yet been devised which answers for use under water-closets; the

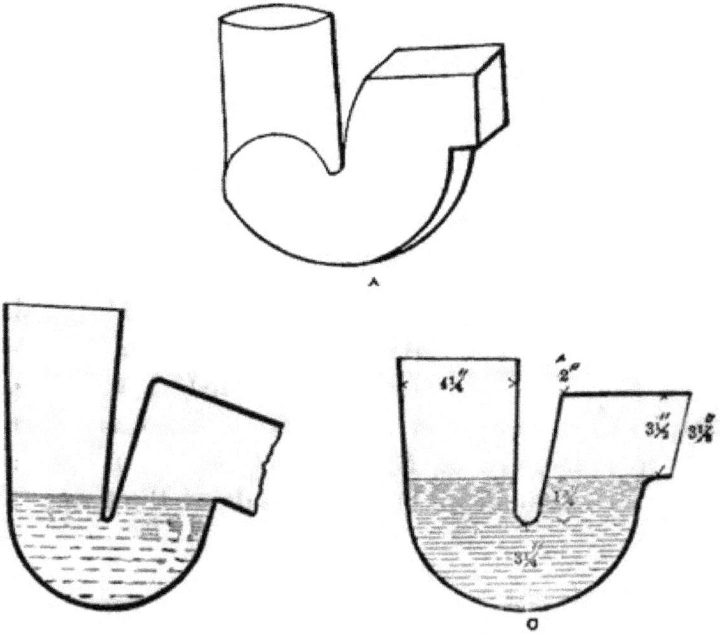

FIG. 93 a, b, c.—Hellyer's Anti-D-Trap.

water-seal traps are the only ones to be relied upon; flap-valves or ball-valves, in connection with water-closet traps, are sure to get out of order after some use.

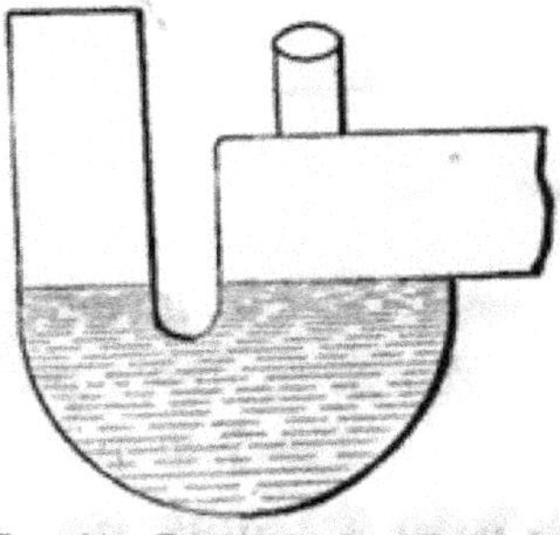

FIG. 94.—Buchan's Anti-D-Trap.

The best traps for use under water closets are the S, $\frac{3}{4}$ S and P traps, made either of lead, iron or earthenware. The last named kind are preferable to any other on account of their cleanliness, but an iron trap may have its inside surface smoothed by *enamelling* the trap. The drawn lead traps, known as "Du Bois" traps (Figs. 96, 97 and 98), are equally smooth on their inside. They are decidedly superior to hand-made lead traps, which, after years, are liable to show defects

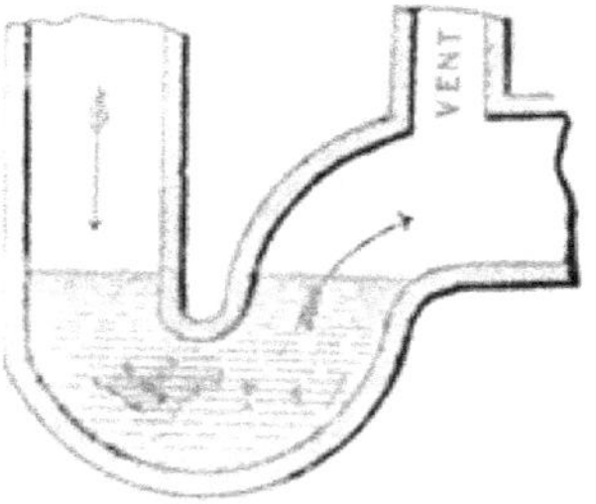

FIG. 96.—Lead P or ½ S-Trap, with vent attached.

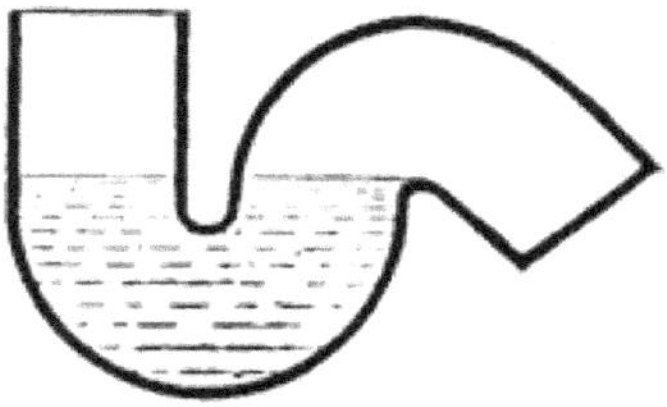

FIG. 97.—Lead ¾ S-Trap.

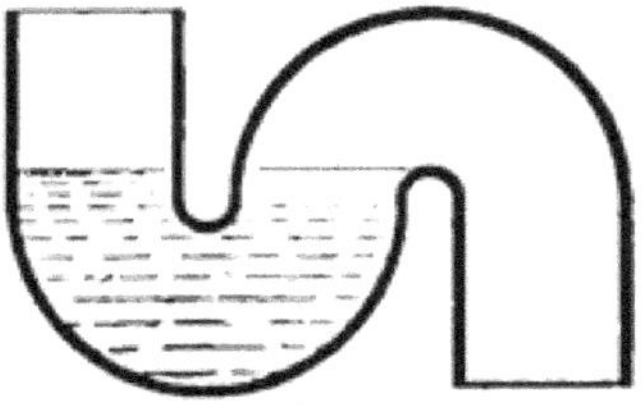

FIG. 98.—Lead S-Trap.

at the seams, and to cast lead traps, which often have sandholes and other defects.

Earthen or porcelain traps for water closets are always set above the floor, iron traps are placed above as well as below the floor, while lead traps are mostly set below the floor, between joists. Since as little plumbing as possible should be hidden from view, it is in most cases preferable to have the trap in plain sight and easy of access. Should the water closet apparatus selected require a trap below the floor, it is much better to use an iron enamelled trap, for a lead trap may have nails driven in at the top by careless carpenters, or may get displaced.

Speaking of water closet traps, I must not forget to call attention to an unsatisfactory manner of *trapping* water closets, by omitting the water seal trap and relying for a barrier against gases solely upon the water in the closet bowl, held in place by a tight-fitting plunger or plug, or else a flap-valve, or slide-valve. Any of these arrangements may get out of order, and the water will then run out of the closet bowl. This shut-off being open, drain air may easily find its way into the house, for a *constant* downdraft from the closet into the soil pipe cannot be depended upon.

Traps for Sinks, Bowls, Tubs, etc.

To choose a proper trap for use under a tub, sink or bowl, is often a rather difficult problem, requiring sound judgment, skill and large experience, since each of the numerous patented devices in the market is, in the opinion of its inventor, the only safe and reliable one to use, or, as it is commonly expressed, "the only positive cure against sewer gas." Each of these patented devices will, upon examination, be found to possess certain merits, which, however, are generally counterbalanced by one or more drawbacks. For instance, one trap may be self-cleansing, but extremely liable to lose its water seal, while another trap may be safe against siphonage or back-pressure, but liable to accumulate grease and filth. One trap may answer under certain conditions and in a certain locality, while in another position another trap might be preferable.

There are bell traps of various descriptions, D

traps, Dip traps, Bottle traps, and various kinds of S traps. All of these have as a barrier against gases a water seal of more or less depth. There are other traps, which have not only such a seal by water, but also a *mechanical appliance* to shut off gases, such as floating balls of rubber or metal; heavy self-seating valves, either rubber or metal balls, or else a conical shaped valve to exclude sewer air. There are also traps provided with flap valves, opening with the current of water and shutting against back-pressure from the soil pipe. Other traps have, in addition to the water seal, a seal of mercury. Finally, a large number of traps and trap attachments have been invented, the construction of which is such as to render siphonage impossible, or at least very difficult.

Many traps of each of the above groups, though sold under different names, are identical in principle and practically the same in construction, so that it often has been a matter of wonder to the author to understand how they all could have been patented as a "new and original invention."

We will now proceed to describe the more common forms of traps of this class.

For sinks, no trap has been used as extensively as the bell-trap (Fig. 99), although probably no other trap offers as little security as this one against sewer gas. It is not self-cleansing, has far too little water seal to resist siphonage, back pressure, and evapora-

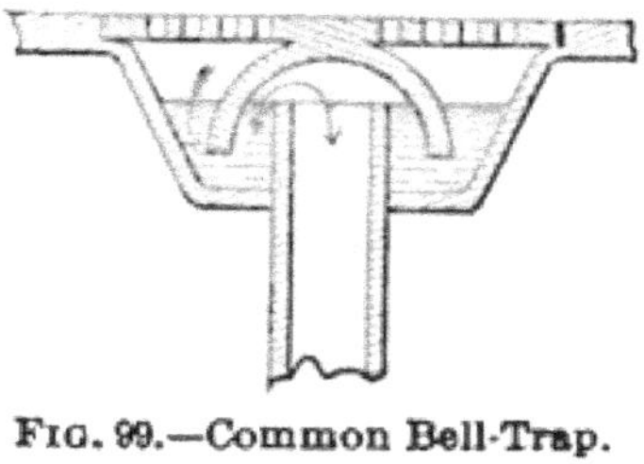

FIG. 99.—Common Bell-Trap.

tion, it gets readily choked, and is generally rendered worthless if servants should remove the top strainer, thus doing away with what little seal the trap possessed.

It will suffice simply to mention the Antill Trap (Fig. 100), which is little better than the bell-trap,

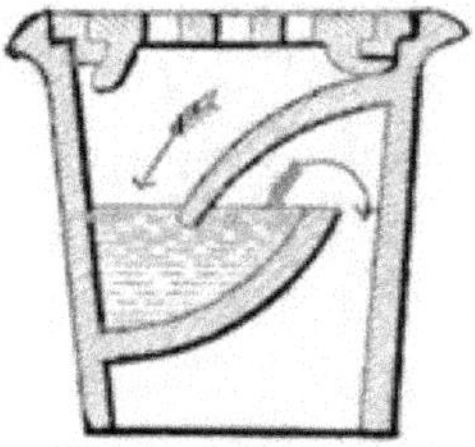

FIG. 100.—Antill's Trap.

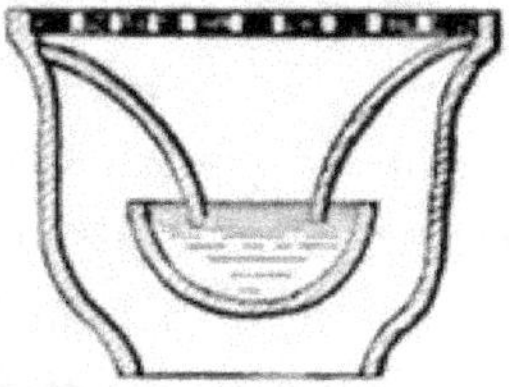

FIG. 101.—Jennings' Improved Bell-Trap.

although it is an improvement upon the latter, in as much as its strainer and dip is not so readily removed.

Fig. 101 shows a vertical section of Jennings' improved bell-trap. It is certainly vastly superior to the common bell-trap; its water seal is not broken by removing the strainer. The stream of water from the sink is concentrated in the inlet pipe, and the inverted bell is consequently better scoured and cleaned; there is also less chance for stoppages, but the upper corners form recesses for the collection of grease, which will decompose. This, however, could easily be remedied by a proper rounding off of the upper corners. This trap is constructed and made by Geo. Jennings, of London, either in lead, with brass grates and bells, or else in galvanized iron.

A good sink trap, made by Messrs. Tye and Andrews, of England, is shown in Fig. 101a. It

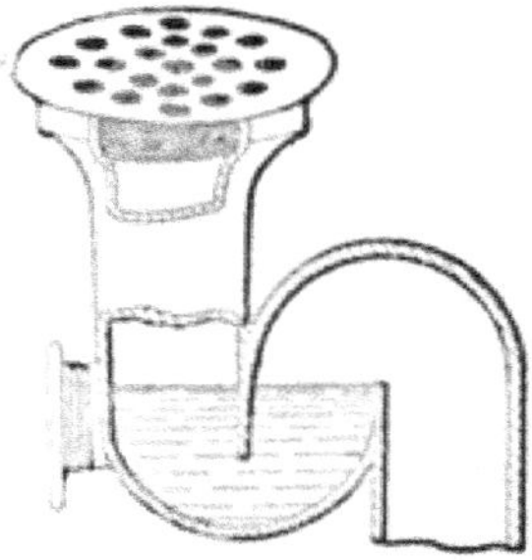

FIG. 101a.—Tye & Andrews' Sink Trap.

is a siphon trap, provided with a trap screw at the side of the trap for cleaning purposes, and with a strainer on top, screwed into the body of the trap, in order to prevent its ready removal by servants, who otherwise would brush all kind of rubbish into the trap, thus causing frequent obstructions.

A small D-trap, such as shown in Fig. 102, is often used under wash-bowls or tubs, but we must earnestly protest against the use of a contrivance so easily fouled. Not even the fact that it is not readily siphoned, can make this trap more acceptable for use in the sewerage of dwellings.

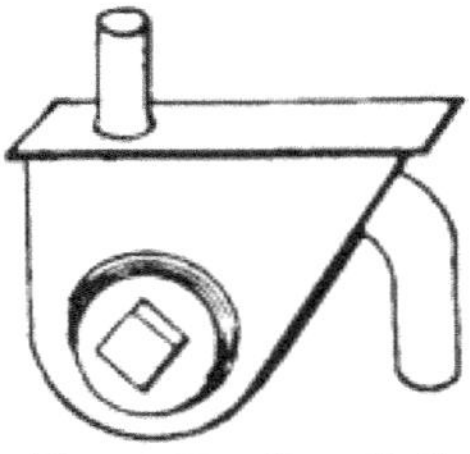

FIG. 102.—Small D-Trap.

There are other traps which have a vertical dip dividing the water chamber in two parts, and thus establishing a water seal. Of these I mention Adee's

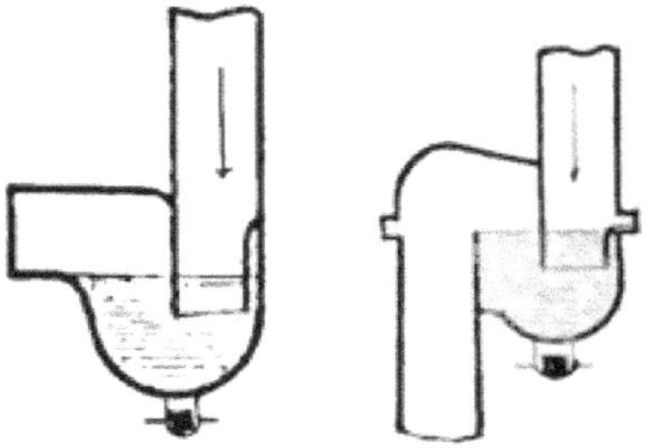

FIG. 103.—Adee's Traps.

P and S stench traps, Fig. 103, Brandeis' paragon trap, Fig. 104, Tylor's trap, Fig. 105, and others.

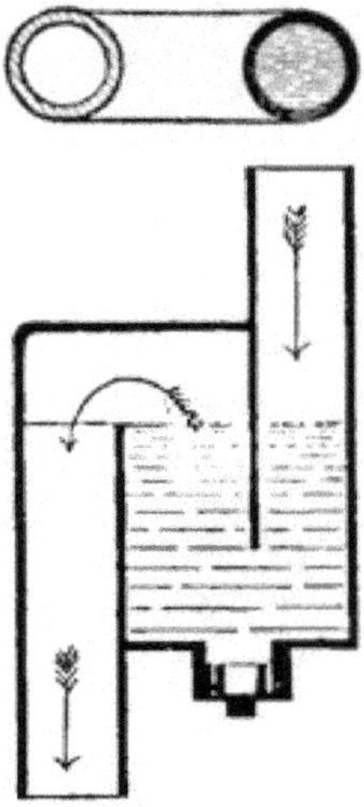

FIG. 104.—Brandeis' Paragon Trap.

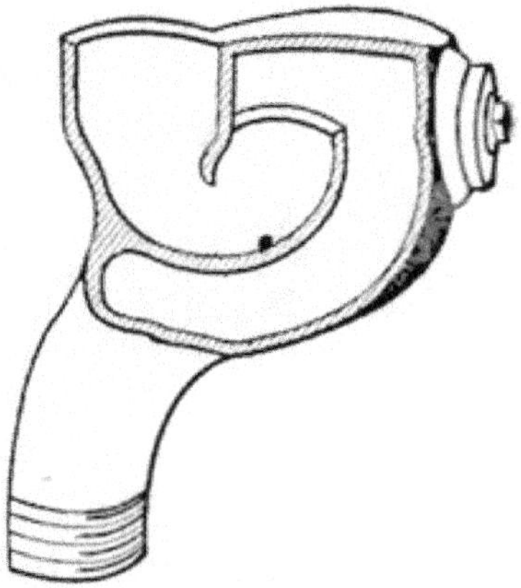

FIG. 105.—J. Tylor's Trap.

Some of these may be more difficult to siphon than the round siphon trap, but they are not self-cleansing traps, and for this reason cannot be recommended for general use.

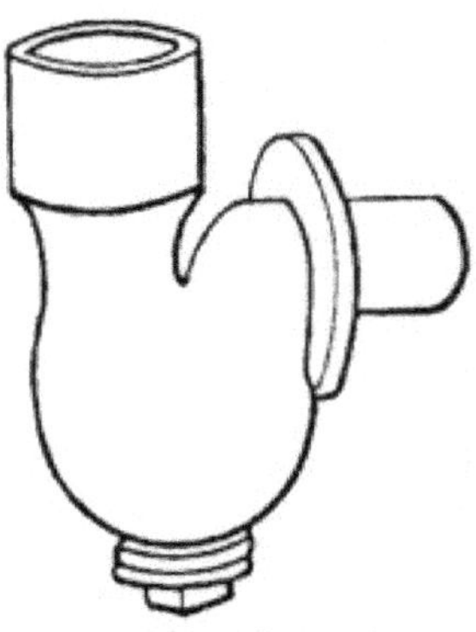

FIG. 106.—J. Tylor's S-Trap.

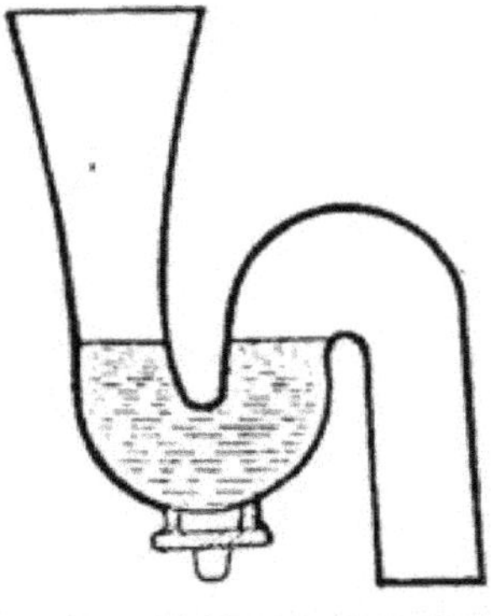

FIG. 107.—Hellyer's Sink S-Trap.

Fig. 106 represents an S-trap for sinks, tubs, urinals or bowls, made in brass by J. Tylor, in

London. Such a trap of brass is preferable to lead or iron traps where fixtures are left entirely exposed to view.

Mr. Hellyer proposes the S-trap, shown in Fig. 107, for use under sinks or other shallow vessels. By enlarging the mouth of the trap at the sink opening, a larger quantity of water drops on to the water level in the trap, which is thus more readily and more effectively flushed.

Of lead siphon traps, the "Du Bois" drawn traps, made by hydraulic pressure in the same manner as lead pipe, possess great advantages over those cast in moulds, or those made with seams by hand. Fig. 108 represents the "Du Bois" S-trap, Fig. 109 his ¾ S-trap, Fig. 110 his P-trap, Fig. 111

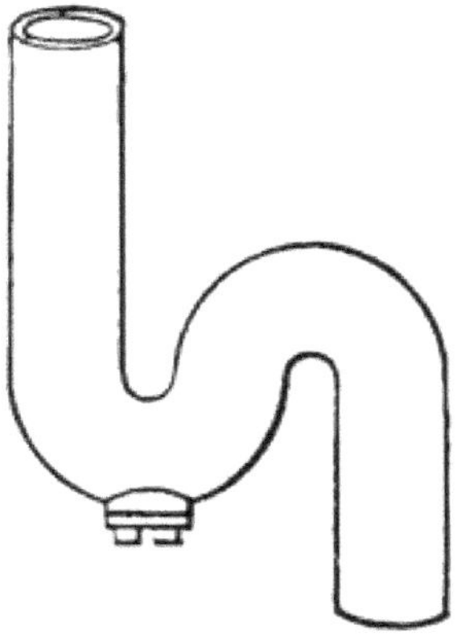

FIG. 108.—Du Bois' S-Trap.

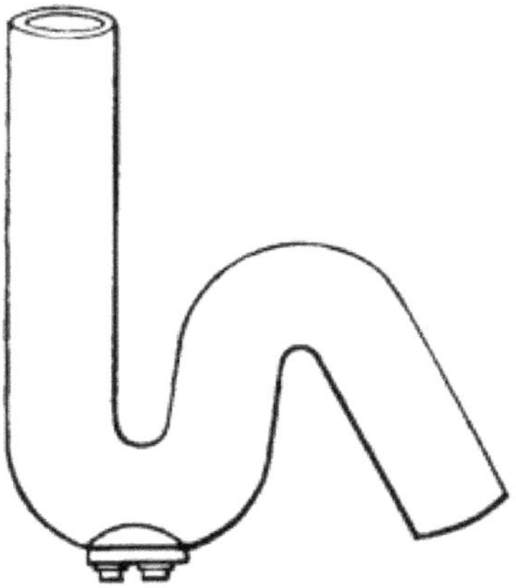

FIG. 109.—Du Bois' ¾ S-Trap.

the running trap, and Fig. 112 the "bag" trap, shaped so as to bring the trap outlet vertically below the inlet.

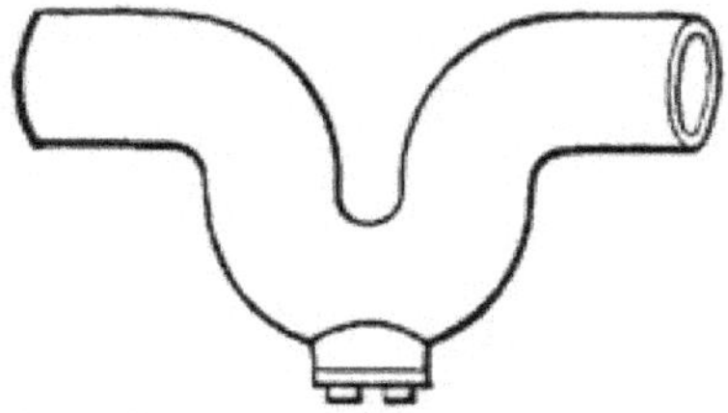

FIG. 111.—Du Bois' Running Trap.

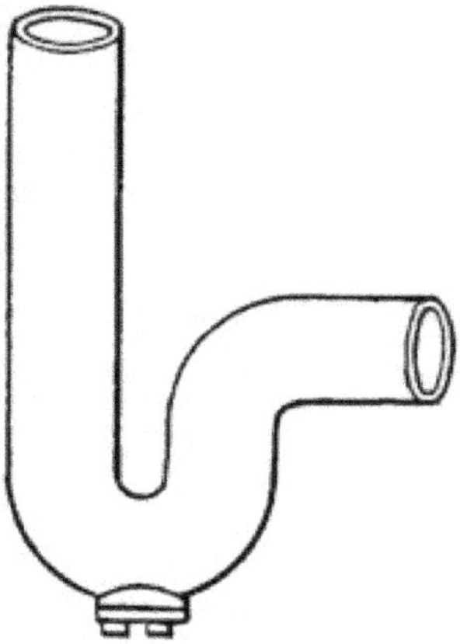

FIG. 110.—Du Bois' P-Trap.

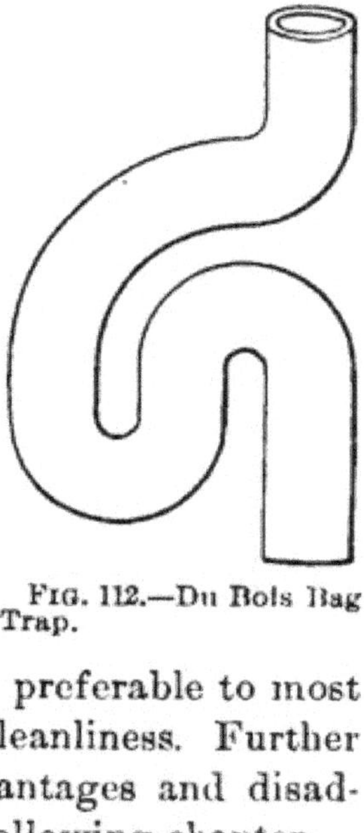

FIG. 112.—Du Bois Bag Trap.

Such lead traps are made of various weights; none but the extra heavy trap, equivalent in weight to the heaviest lead waste pipe, should be used in good plumbing.

Such S-traps are, as a rule, preferable to most other traps on account of their cleanliness. Further reference to S-traps, their advantages and disadvantages, will be made in the following chapter.

Another group of traps for sinks, bowls or tubs, may be called *bottle traps*, the general shape of the trap being somewhat like a bottle, with an inlet pipe in the centre and an outlet pipe on the circumference of the bottle. Figs. 113 and 114 represent Adee's bottle traps, in half-S and S-shape.

Fig. 115 shows a trap used extensively in Bos-

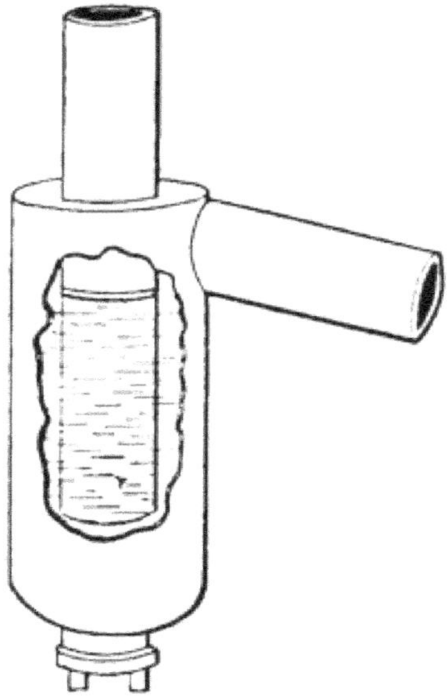

FIG. 113.—Adee's Bottle Trap, half S-shape.

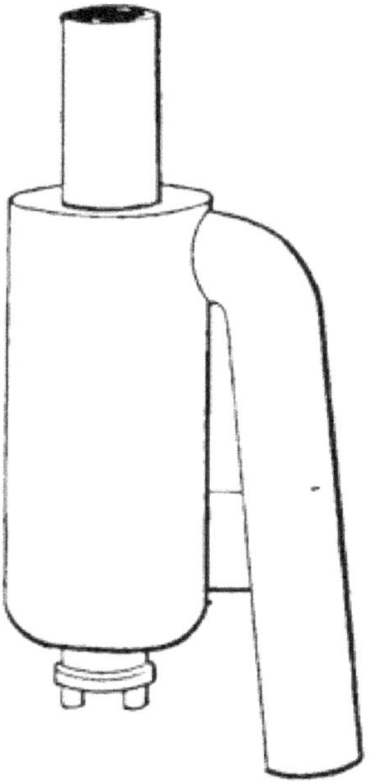

FIG. 114.—Adee's Bottle Trap, S-shape.

ton plumbing work, and called the round trap. With a flat bottom it soon accumulates deposits, as

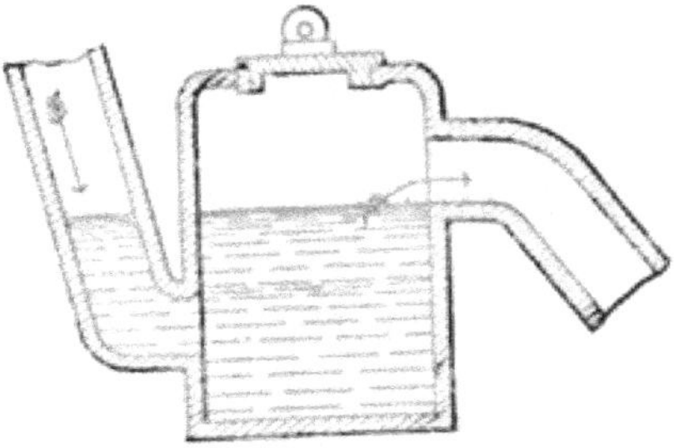

FIG. 115.—Boston Round Trap.

FIG. 116. — Boston Round Trap, with filth accumulation.

shown in Fig. 116. It may be somewhat improved by rounding off the bottom, as shown in Fig. 117. Bottle traps are often safe against siphonage,

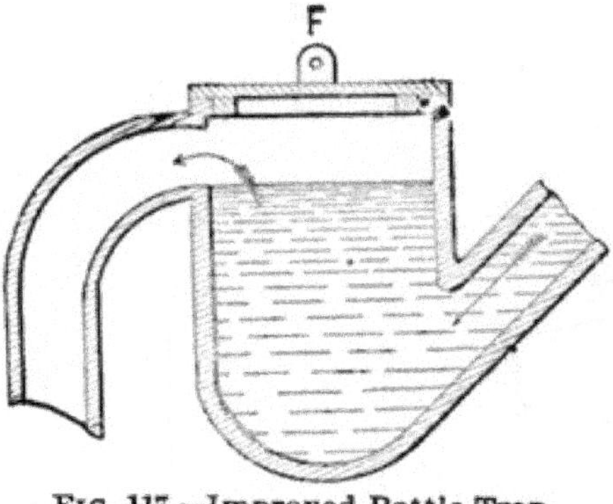

FIG. 117.—Improved Bottle Trap.

where an S-trap would lose its water seal, but they are not self-cleansing; an ordinary discharge from a sink or bowl will not entirely change its contents, and after collecting filth, the bottle trap may not be much safer against siphonage than the S-trap.

Fig. 118 shows another reservoir, or bottle trap,

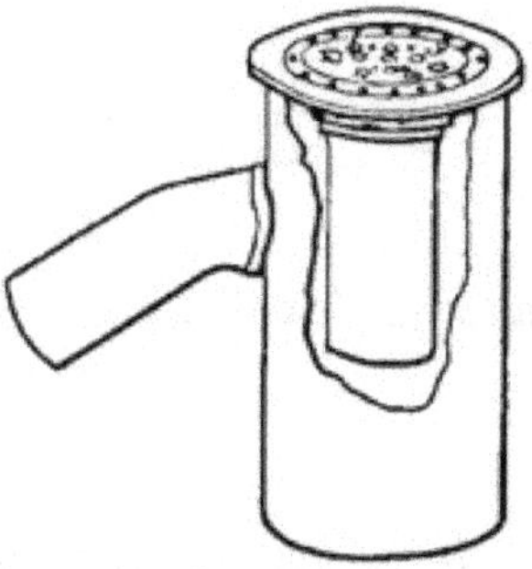

FIG. 118.—Bolding's Patent Bottle Trap.

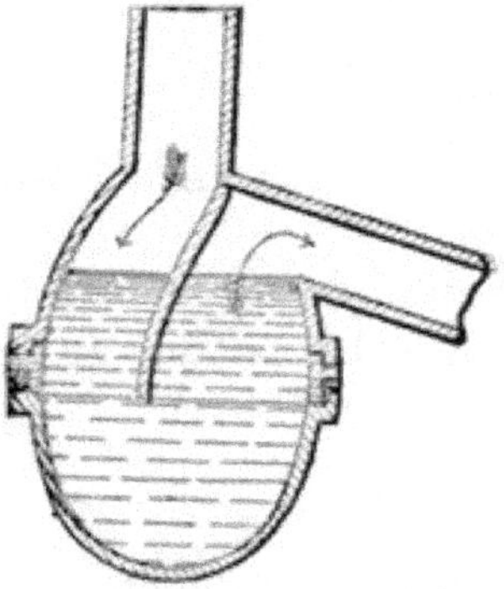

FIG. 119.—Brandeis' "Climax" Trap.

made for sinks, by John Bolding, in London, which needs no further explanation.

Fig. 119 illustrates Brandeis' "Climax" trap, which is a bottle trap with quite a large dip, and has a cup at the bottom, which can be unscrewed for cleaning purposes.

Fig. 120 represents Stidder's patent soap trap, which also belongs to the group of bottle traps, as well as Buchan's round bottle trap, Fig. 121.

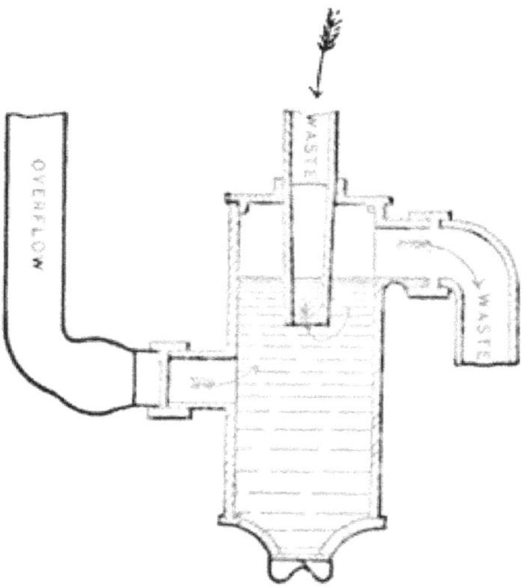

FIG. 120.—Stidder's Patent Soap Trap.

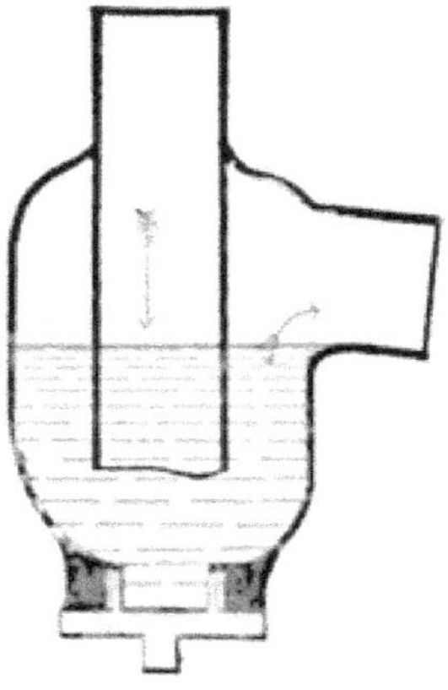

FIG. 121. — Buchan's Round Bottle Trap.

Fig. 122 shows a trap made by Claughton, in England, of a round or oval section, as shown, with a cleaning screw placed at one side, and having two inlets, one for the waste, the other for the overflow of a basin, or sink. The outlet is brought directly under the inlet, which is of advantage, as far as appearance is concerned.

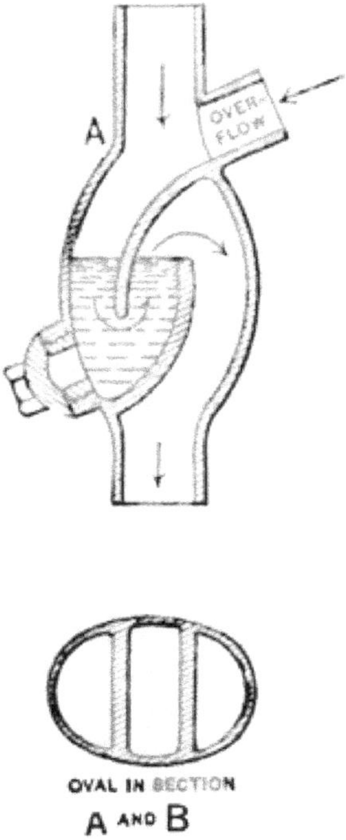

FIG. 122. — Claughton's Trap.

Fig. 123 is another style of trap, made by the same manufacturer, and Fig. 124 shows his sink trap, which is very similar in appearance to the "Climax" trap, described above. Instead of a removable cup, the trap is provided with a brass cleaning screw.

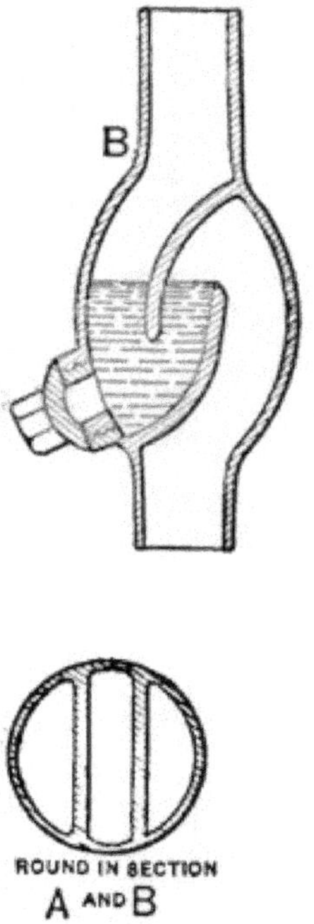

FIG. 123.—Claughton's Trap, for sinks.

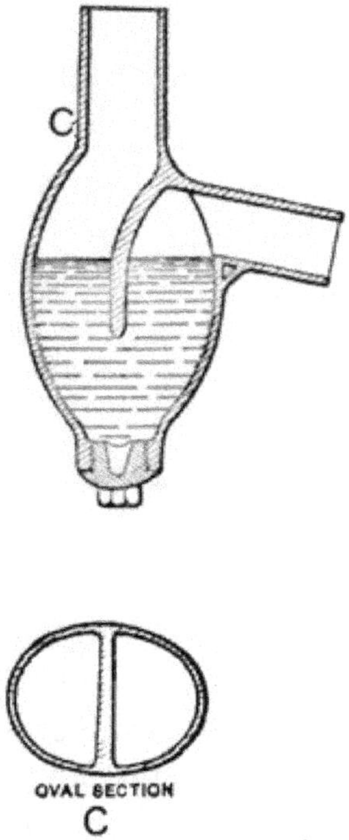

FIG. 124.—Claughton's Trap, for sinks.

Another recent trap is Connolly's "Globe" trap, which is made in various styles and of various materials, either of copper, with brass connections, or of sheet lead, or of glass, with brass connection. The traps are made either in S or P-shape, or as running traps. Fig. 125 illustrates the P-shaped globe trap, of sheet lead; Fig. 126 the running globe trap, of copper; Fig. 127 the glass globe trap, with brass connection.

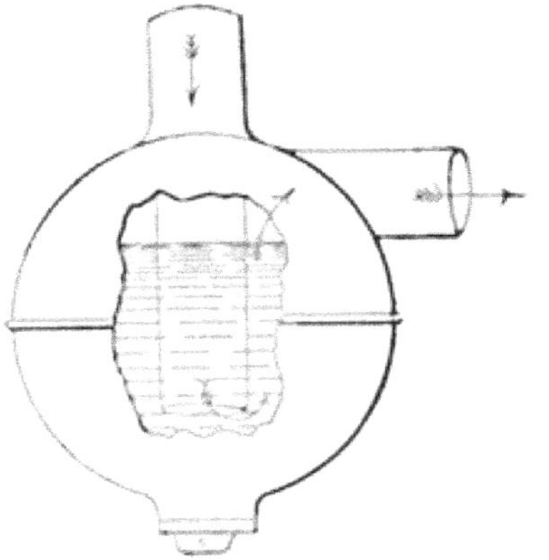

Fig. 125. — Connolly's Patent Globe Trap, of sheet lead, P-shaped.

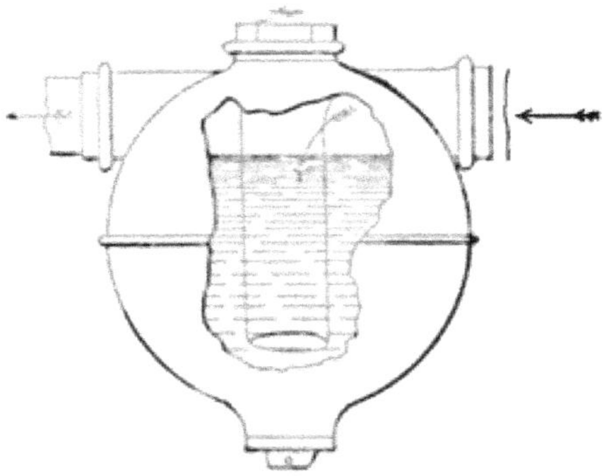

Fig. 126. — Connolly's Patent Globe Trap, of copper, running trap.

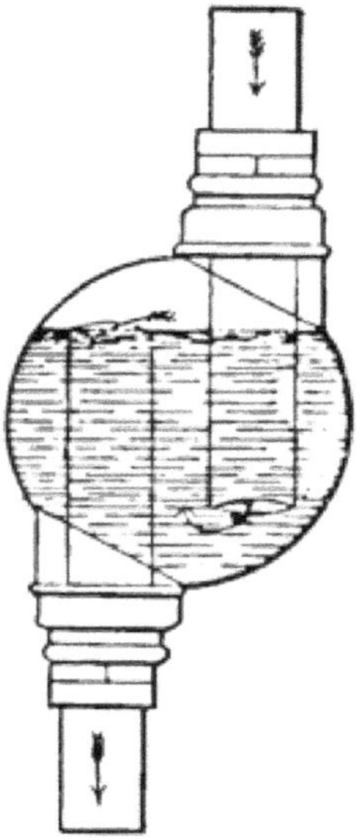

Fig. 127. — Connolly's Glass Globe Trap, with brass connections.

The application of the glass globe trap to wash-bowl waste and overflow is shown in the sketch, Fig. 128. There is, undoubtedly, a certain advantage in having a globe made of glass, which enables any one to see at a glance whether the seal in the trap is destroyed or not. I am unable, without repeated actual experiments, to verify the claim of the manufacturer, that the globe trap cannot be rendered inefficient by siphonage. The construction of the trap is such as to allow the unscrewing of tbe overflow pipe, in case this pipe should need cleaning.

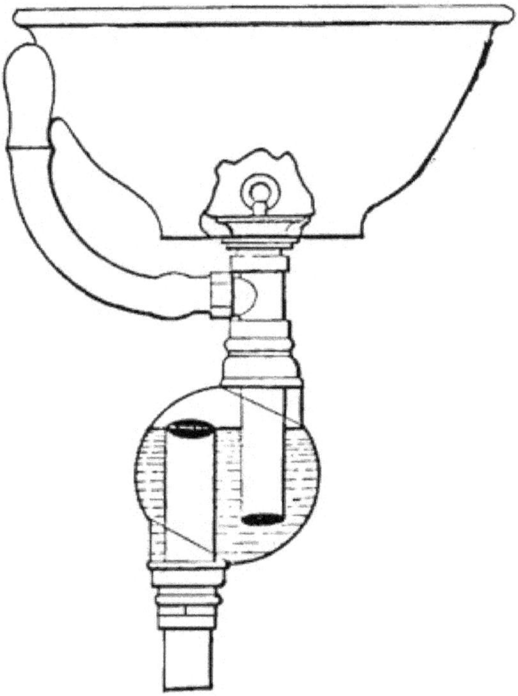

Fig. 128.—Washbowl, with Connolly's Glass Trap.

Another late invention is Pietsch's stench trap, Fig. 129. It is a bottle trap, with deep seal, which is not easily siphoned. The inlet pipe from sink, or bowl, is provided above the water seal with a flap-valve, which, under ordinary circumstances, is supposed to shut tightly, and to be still more firmly pressed against its seat by back-pressure. Its object is the prevention of siphonage by admitting air from the fixture to the waste pipe, in case a strong

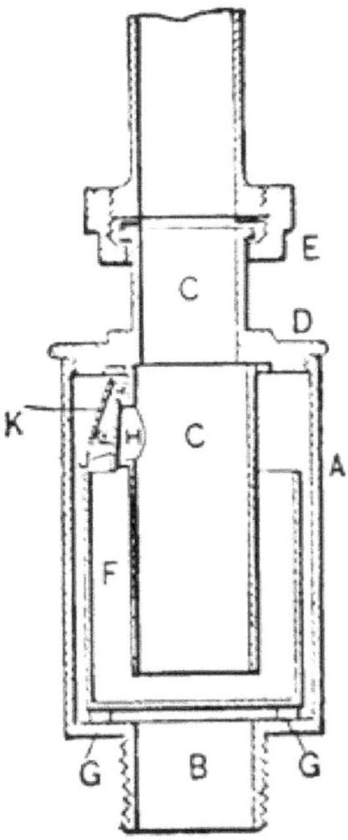

FIG. 129.—Pletsch's Stench Trap.

suction should be applied to the water in the trap. The danger with this trap lies in the flap-valve, which may get out of order and will not tightly shut, in which case gases could freely pass from the soil or waste pipe up the fixtures into the rooms, in spite of the deep water seal.

We must now consider a few of the more important *mechanical* traps used in plumbing work. The additional mechanical seal by a valve, ball or flap, is intended to give increased security in case of back-pressure, and in case of evaporation of the water in the trap, or in case of the water being removed by siphonage, they form a seal, which, with the gravity valves, will depend upon the accuracy with which the seat is turned. In the case of traps with a floating ball, the seal is preserved only as long as the water is not lowered so much as to drop the ball from the mouth of the inlet pipe.

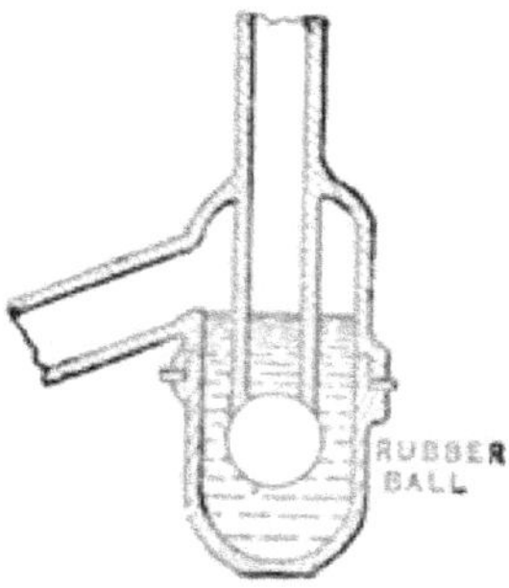

FIG. 130.—Bower's Trap in section.

Among mechanical traps which have been used very extensively of late, I mention the Bower trap, shown in section, in Fig. 130; in Fig. 131,

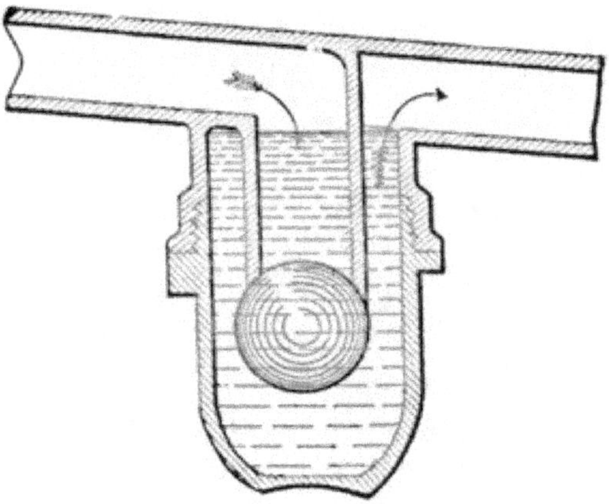

FIG. 131.—Bower's Running Trap, in section.

which represents a running trap, and in Fig. 132, which shows a wash-bowl trapped by this trap, and Fig. 132a, showing, on a large scale, a view of the trap. Its construction and action has been thus described by the Committee on Science and Arts of the Franklin Institute:

"The invention consists in providing a sewer gas trap with a floating valve which will permit the flow of water and gases carried with the water in one direction, and prevent their regurgitation. The inlet pipe of the trap extends downward into a chamber, which is of somewhat larger dimensions than the inlet pipe. The outlet pipe is arranged so that its discharge opening is relatively such to the lower end of the inlet pipe that the level of the water in the trap is always a considerable distance above the opening of the inlet pipe, and the trap is ordinarily sealed by water. A float ball or valve (preferably a hollow ball of rubber) is placed in the trap beneath the end of the inlet pipe, and this valve is constantly immersed in the liquid—the dimensions of the trap being such that the ball cannot escape upwards alongside of the inlet pipe.

FIG. 132.—Washbowl, trapped by Bower's Trap.

When water is poured into the inlet of the trap, the ball is forced away and permits a free passage to the outlet.

Should anything occur to bring a pressure upwards from the outlet of the trap, the ball (already held in its place by floatation) is more firmly pressed into the seat, and prevents the passage of liquids or gases through the trap from its outlet to its inlet. The advantages possessed by this device above others with check valves consist in the constant approximation of the valve to its seat, and the ease and little force with which it is displaced and replaced when water has passed the trap.

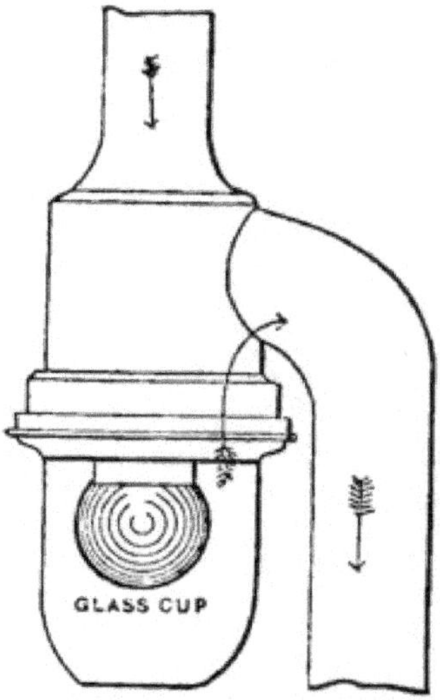

FIG. 132a—View of Bower's Trap.

As an essential to the proper working of the device, care must be taken to select material for the valve that shall secure floatation; if a hollow rubber ball be used, it must be perfectly airtight. . . . ."

Amongst the advantages of this trap I mention the following: its seal is not broken by evaporation, nor by back-pressure; it affords a seal against absorbed gases and against back water; the cup is removable, giving access to all parts of the trap; it may be fitted with glass cup, exposing the rubber ball and the water seal; freezing will rarely, if ever, injure the trap, as the hollow rubber ball may be sufficiently compressed to allow for expansion; the screw-joint between cup and body of the trap is *below the water-line*, consequently there can be no leakage of sewer gas at this point. This is all-important, and we shall see later that most of the gravity valves do not possess this advantage.

Another merit of the floating valve consists in its lesser resistance to the flow of water from the fixture than that of gravity or flap valves. Finally, I mention the fact that the seal of the Bower trap is not easily lost by siphonage, provided the main soil and waste pipe system has ample ventilation. This fact the author has ascertained while making numerous experiments on the siphonage of traps for the National Board of Health, under direction of Col. Geo. E. Waring, Jr.

With a soil pipe, closed at the top, the author succeeded in removing, by suction, sufficient water from the trap to cause the ball to drop from the inlet pipe, but under all *ordinary combination of discharges* from fixtures, the writer found it impossible to do so, with a soil pipe open at the top and having a fresh air pipe at the foot. I am aware that other writers claim that even then the seal of the Bower trap may be lost by siphonage.

The chief objection to the Bower trap, which I know, is its liability to become filthy in the upper corners. It is quite true that the ball-valve is cleaned at each discharge by being made to revolve in the chamber. It must also be admitted that the ball-valve produces an eddy which assists in scouring the bottom of the trap. But the upper corners do not get any benefit from this scour, while, especially, if used under sinks, grease will collect and remain there.

The author offers, in Figs. 133, 134 and 135, some suggestions for improving the shape of this trap so as to render it more self-cleansing while retaining the advantages of the original trap.

A floating ball-valve, somewhat different and less useful that the Bower trap, is Putzrath's back-

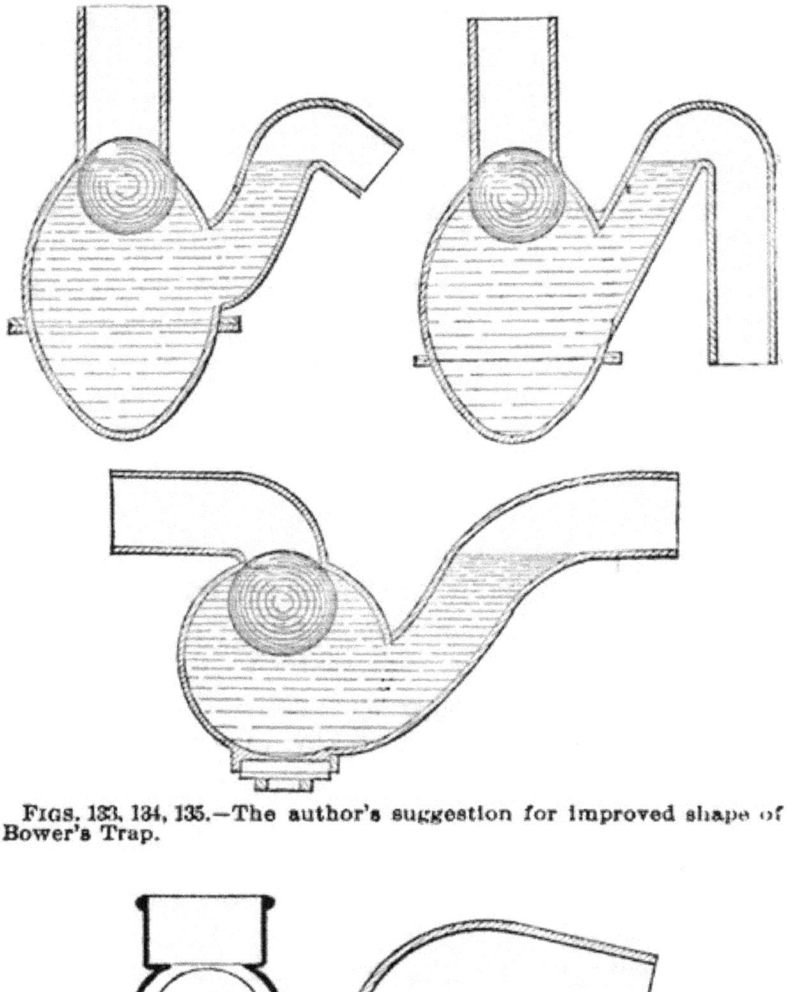

FIGS. 133, 134, 135.—The author's suggestion for improved shape of Bower's Trap.

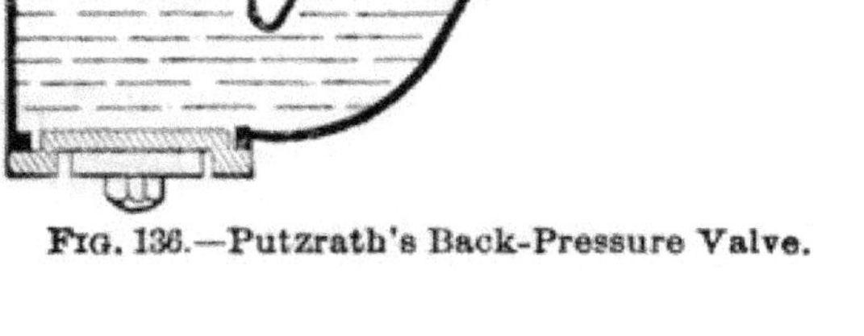

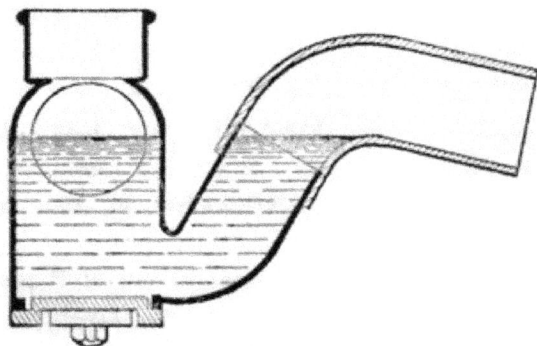

FIG. 136.—Putzrath's Back-Pressure Valve.

pressure valve, Figs. 136 and 137. It is made in Germany for use under sinks, tubs, bowls and water-closets, but it appears to me that for the latter purpose, the trap is wholly unfit. It would soon choke and become obstructed, and the rubber ball would soon be destroyed. As a back-pressure valve, the trap may be quite efficient.

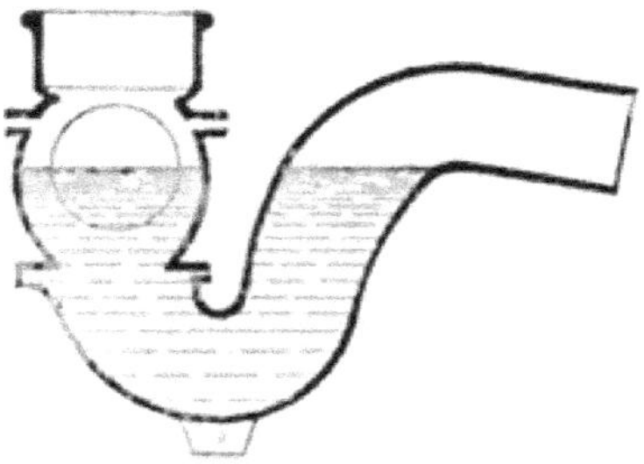

FIG. 137.—Putzrath's Back-Pressure Valve.

Fig. 138 shows a rather complicated form of

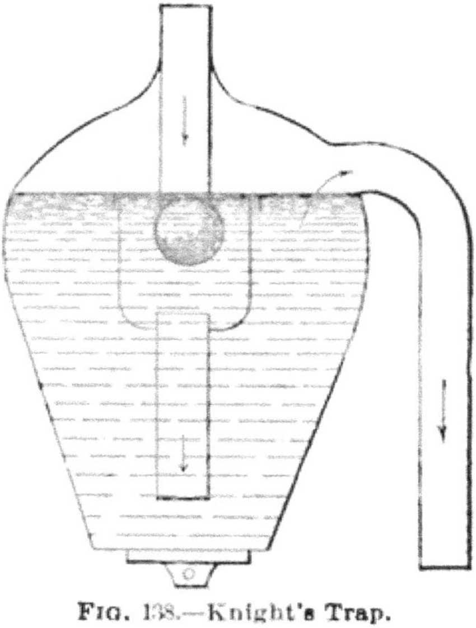

FIG. 138.—Knight's Trap.

stench trap, the invention of Thos. G. Knight, of Brooklyn. It is a trap with a floating ball, which is expected to answer the double purpose of sealing against the inlet pipe in case of back-pressure, and against the seat at the outlet, in the case of siphonage, thus preventing the escape of water and leaving the seal in the trap unbroken. The idea of the trap is a good one, but its shape is rather clumsy and much too large, while the trap itself is a reservoir for filth.

We will now briefly consider some gravity valve traps. One of the earlier traps of this group is Waring's sewer gas check valve, Fig. 139. The objection to this, as well as to other gravity valves, is, that they catch hair, lint, etc., especially at the valve seat. The valve will then shut imperfectly and render the mechanical seal useless. For this reason the inventor himself has abandoned the use of this trap, except for waste pipes through which *clean* water only flows, for instance, overflows from tanks.

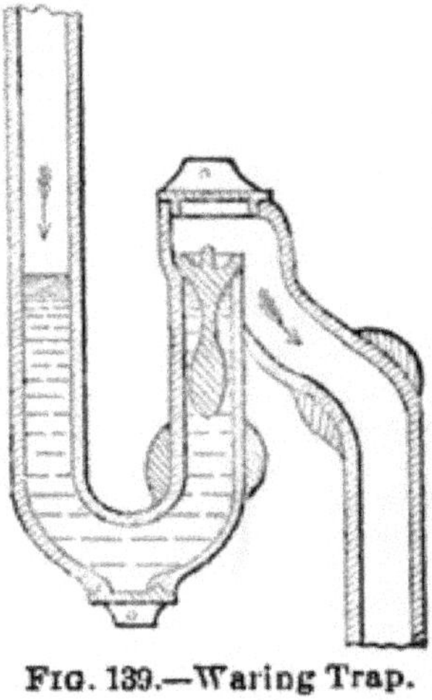

FIG. 139.—Waring Trap.

A round ball of heavy rubber, or metal, is more apt to keep itself and the valve seat clean by revolving. Such a valve is used in the well-known Cudell sewer gas trap, illustrated in Fig. 140. It is made in the S, half S, and running-trap shape, with an enlarged chamber containing the ball valve. In case of siphonage or evaporation, it is claimed that the sinking ball will efficiently keep out by its downward pressure on its seat any sewer gas. The trap is provided on top of the chamber with a removable cover, for cleaning purposes.

The danger with a cover arranged in this manner consists in its being on the sewer side of both the water and mechanical seal. Any imperfection in the joint would render the double seal perfectly useless. I also object to the shape of the Cudell trap, which forms corners and recesses between the inner chamber and the outer walls of the trap, where grease and filth may lodge. This objection could easily be overcome by giving the Cudell trap the shape shown in Fig. 144 or Fig. 145, both of

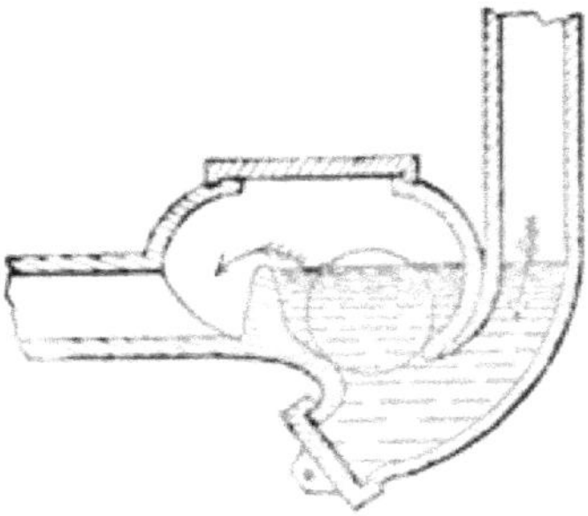

FIG. 140.—Cudell Trap.

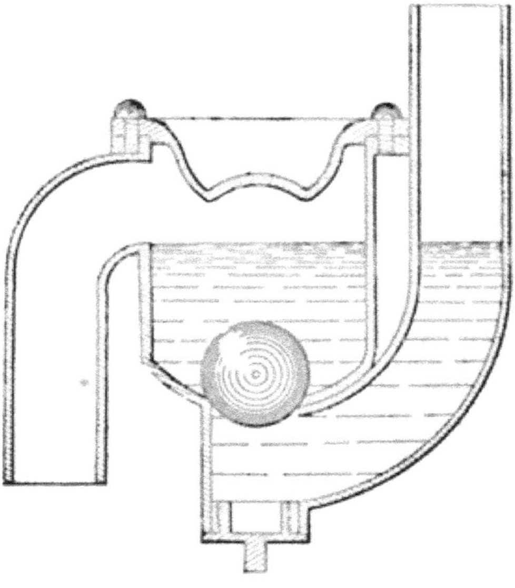

FIG. 141.—Garland Trap.

which sketches represent the author's suggestions for a gravity ball trap.

Almost identical with the Cudell trap are three other traps, namely, the Garland trap, Fig. 141, Buchan's trap, Fig. 142, and Turner's trap, Fig. 143. Fig. 141a illustrates a washbowl trapped by a Garland trap.

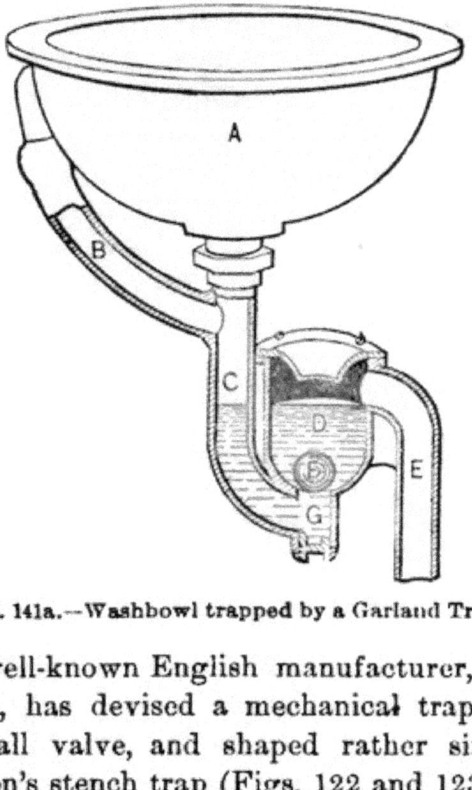

FIG. 141a.—Washbowl trapped by a Garland Trap.

The well-known English manufacturer, George Jennings, has devised a mechanical trap, with a heavy ball valve, and shaped rather similar to Claughton's stench trap (Figs. 122 and 123). Jennings' trap, Fig. 146, was at first made with a metallic seat, but the patentee himself fearing uncertainty of action, never offered it to the public. Now the trap is provided with an india-rubber seating, and the heavy ball shuts off tightly. Should the water in the trap be lost by evaporation or siphonage, the mechanical seal will still, it is claimed, keep out any gases from the waste pipes.

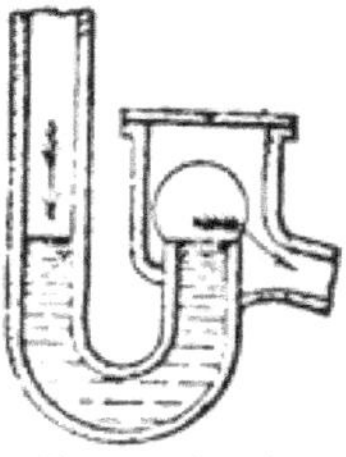

FIG. 142.—Buchan's Trap.

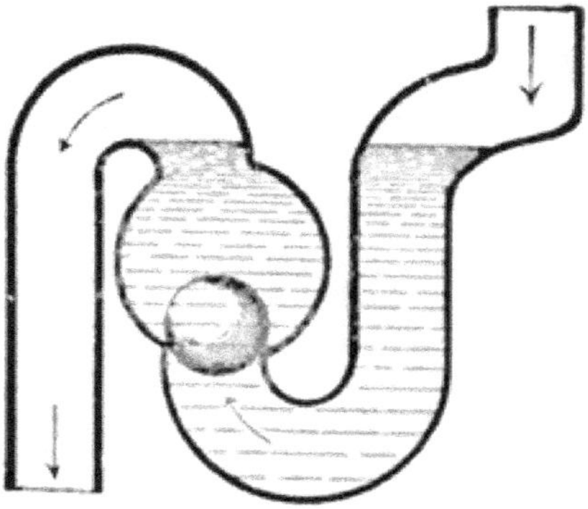

FIG. 143.—Turner's Trap.

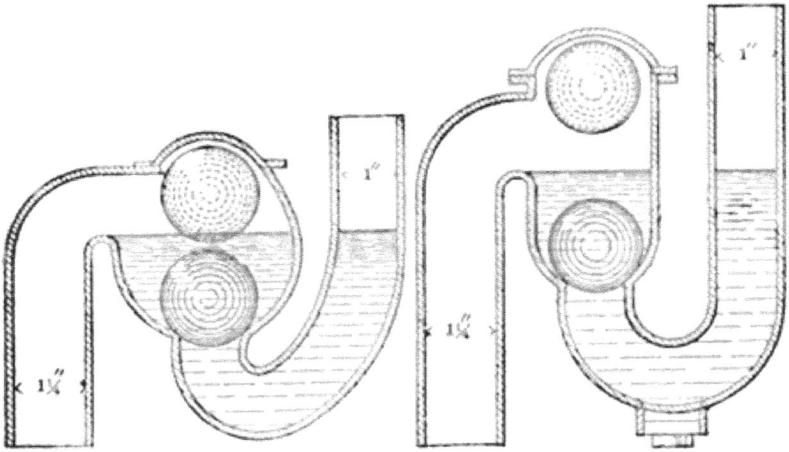

FIGS. 144 and 145.—The author's suggestion for improved shape of Cudell's Trap.

The Bennor siphon trap, Fig. 147, is very similar to the Garland trap, and needs no further description.

Another group of mechanical traps are those with flap valves. Fig. 148 shows Clement's patent trap, made entirely of brass, with a vent pipe attachment, wherever an air pipe is required, and with a cleaning hole, closed tightly by a cover, and

FIG. 146.—Jennings' Trap.

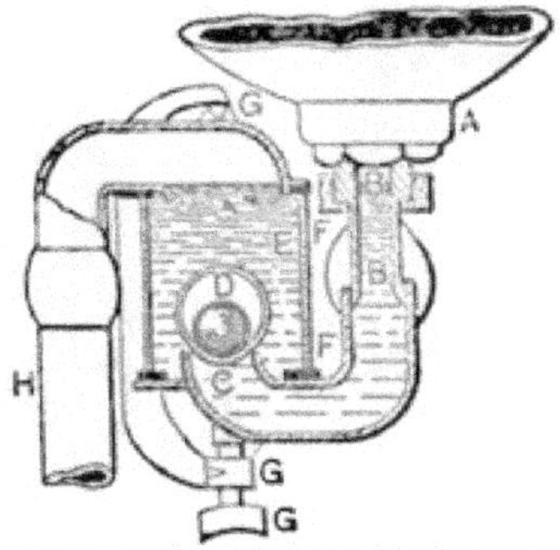

FIG. 147—Bennor's Siphon Trap.

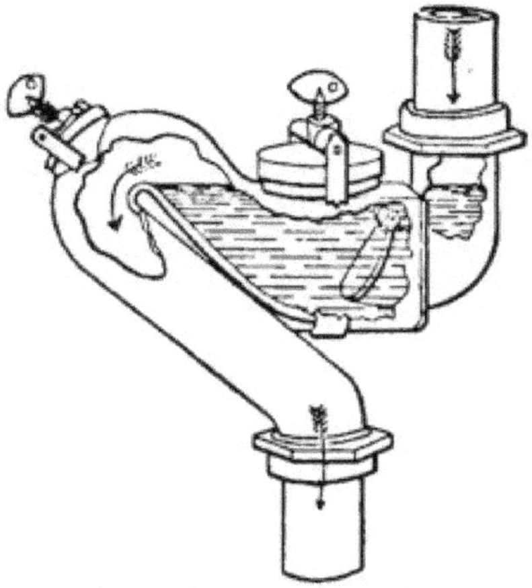

FIG. 148—Clement's Trap.

being below the water line, which is important, as the fact of the cover not shutting tightly is shown by leakage of water.

Stidder's flap valve trap is shown in Fig. 149. It

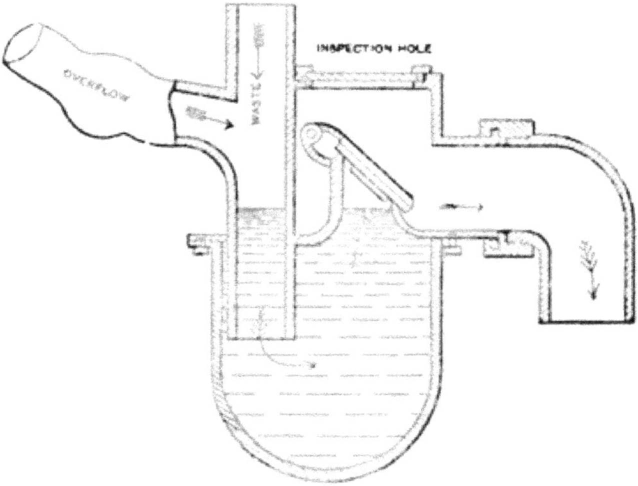

FIG. 149.—Stidder's Patent Flap Valve Trap.

is a bottle trap, provided for additional security against gases, with a flap valve at the outlet.

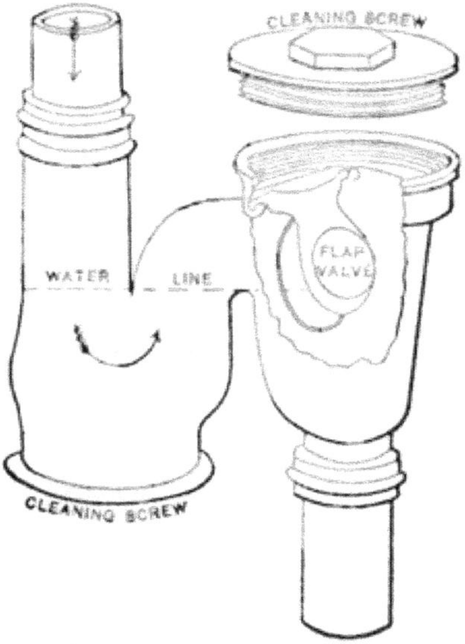

FIG. 150.—Barrott's Trap.

Barrett's trap is illustrated in Fig. 150. Here the flap hangs beyond the water seal, and the trap itself is shaped much like an S-trap.

A further group of traps are the *mercury-sealed traps*, some of these having, in addition to a water seal, a seal formed by a cup resting with its edges in mercury. The cup is lifted at each discharge, and allows the water to pass to the outlet, and when the flow ceases, the cup drops back and forms a seal. Should the water seal be lost in these traps, there is still the mercury-seal remaining. Of such traps, I mention Nicholson's mercury seal trap, Fig. 151, and Spratt's mercury seal trap, Fig. 152. Cohen's trap, Fig. 153, is a mercury seal trap, pat-

Fig. 151.—Nicholson's Mercury Seal Trap.

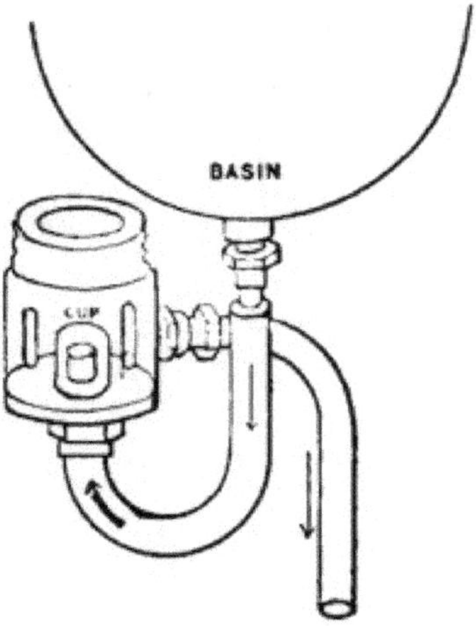

Fig. 152.—Spratt's Mercury seal Trap.

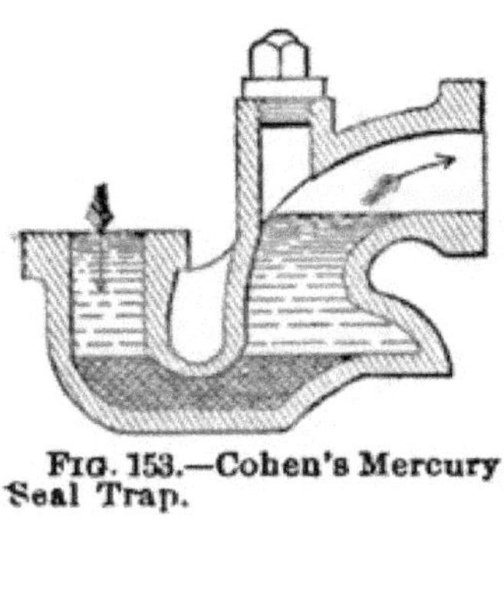

Fig. 153.—Cohen's Mercury Seal Trap.

ented in Germany. It holds in its lowest part just sufficient mercury to form a seal, and if waste water is discharged, the head of water forces the mercury into an enlargement of the outer leg of the trap, until equilibrium is restored. This trap is made in glass or in earthenware, and has a brass cleaning stopper. Another mercury seal trap, introduced quite recently, is Edward's "metallic bar" trap.

A large number of traps and trap attachments have been devised, with the special object of preventing the siphoning out of the water in the trap. I will mention a few of these. Randolph's trap, Fig. 154, has a double gravity valve, its shape being otherwise very much like the Paragon trap (Fig. 104). A glass in the upper side of the trap enables one to inspect the working of the ball valves, which is as follows: When in rest there is, in addition to the water seal, a mechanical seal by the ball, which is half immersed in water. The second ball valve at the outlet also shuts off by its weight, but in case of undue pressure, this would tend to lift the ball. If a discharge occurs, it lifts the first ball, leaving around it a water way through which the water flows out. In rising, the first ball touches the second ball, which is also lifted, to allow the water to pass freely. As soon as the discharge ceases, both

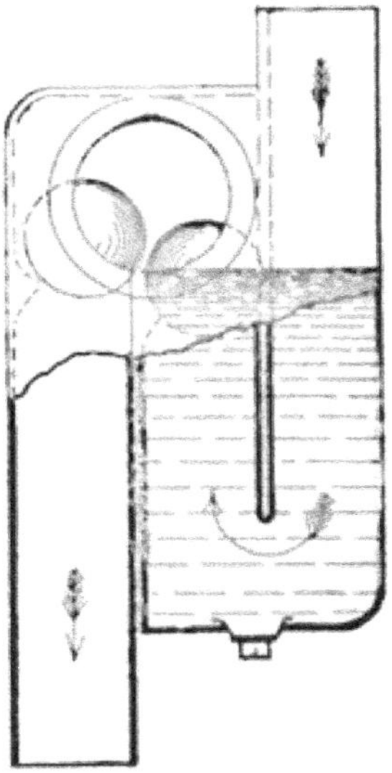

FIG. 154.—Randolph's Trap.

valves drop back into their seat. If a suction should occur, the valve nearest the outlet will efficiently prevent any loss of water by siphonage. The trap is safe against back-pressure, and absorption of gases is rendered almost, if not entirely, impossible. Whether this trap is self-cleansing or not, I am unable to say, without having watched it under continuous use.

Fig. 155 illustrates a vertical section of Morey's trap attachment. "It is soldered on traps already in use at the highest part of the bend. Any tendency of the water passing through the discharge pipe to create a vacuum causes the valve to lift, and the air rushes into the pipe, destroying the vacuum and preventing the trap being drawn dry. The suction ceasing, the valve drops by its gravity into its seat, forming an air-tight joint, preventing the escape of noxious vapors."

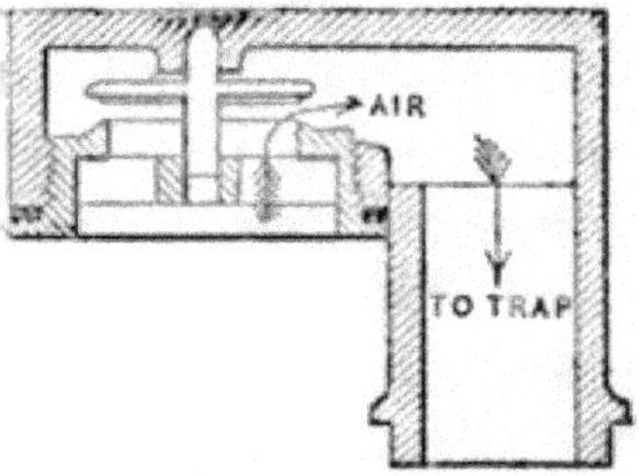

FIG. 155.—Morey's Trap Attachment

Fig. 156 shows Scarborough's trap, which is said to be non-siphoning. It is an S-trap, on which is placed an air chamber between its upper and lower bends. Should a discharge occur filling the full bore of the waste pipe, there is a tendency of the water to drop from the crown of the trap both ways, that is, into the outlet and back into the body of the trap. There is, consequently, a ten-

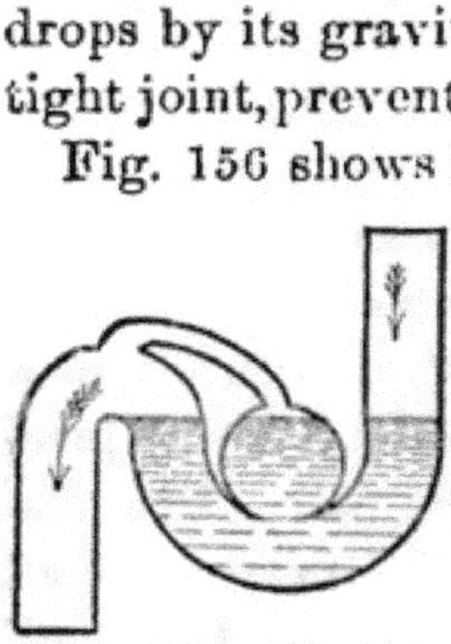
FIG 156. — W. Scarborough's Trap.

dency to a vacuum at the crown of the trap. As this is connected to the air chamber, water is sucked up into it. When the suction ceases, this water drops back into the body of the trap to form a seal. The air chamber should be so proportioned as to hold at least a quantity of water equal to that in the trap. Such a device may work well for clean water, but with soapy or greasy discharges, it will soon get clogged.

Fig. 157 illustrates the author's suggestion for

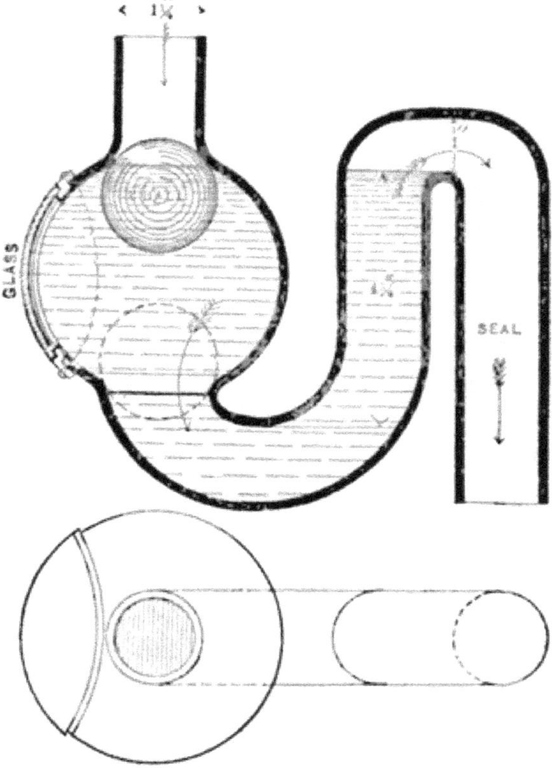

FIG. 157.—The author's suggestion for a Non-Siphoning, Self-Cleansing Trap.

a non-siphoning trap. It is a water seal trap, provided with a double-acting ball valve. It may be made of brass or copper, and consists of an S-trap, having at the house-side of the dip a globular enlargement. There is a removable section of glass to inspect the working of the trap, and to remove the ball in case it should get dirty or water-soaked. The floating ball may be made of india-rubber. As it appears under a fixture not just used, it is really nothing but a Bower's trap, of improved shape. In case of a discharge, every part of the enlarged chamber receives a washing and scouring; and moreover, that part of the trap having an S-shape, will certainly always keep clean. The ball is revolved by the eddies when water flows through the trap and increases the scouring action of the flow. Now, suppose a siphonage should occur, either from a discharge of the fitting itself, or from a discharge through the main soil pipe. The suction will remove some water from this trap until the light ball drops into the position shown by dotted lines. There should be a neatly-turned seat to receive the ball in order to have it shut tightly. Then no more water can be sucked out, and the ascending leg of the S-trap will remain full of water. When the suction ceases, the water column drops back and forms a sufficient water seal, the rubber ball floats on the water, and as soon as the next discharge from the fitting occurs, is brought up to its seat. Until then the trap is simply a water-sealed trap; but no matter how often a suction may occur, it will not remove the quantity of water left in the trap, equivalent in capacity to the contents of the ascending leg.

Two more non-siphoning traps close the list, which makes no pretension at all to be complete. Fig. 158 is a common S-trap, furnished at its

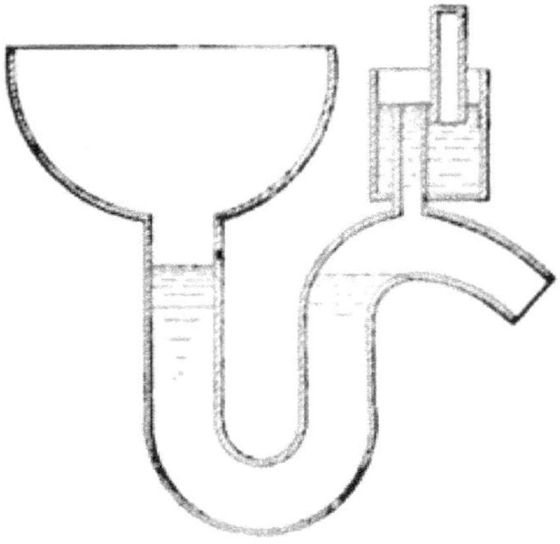

FIG. 158.—Pettenkofer's Trap Attachment.

crown with Pettenkofer's trap attachment, which prevents the siphoning of the S-trap, but is soon

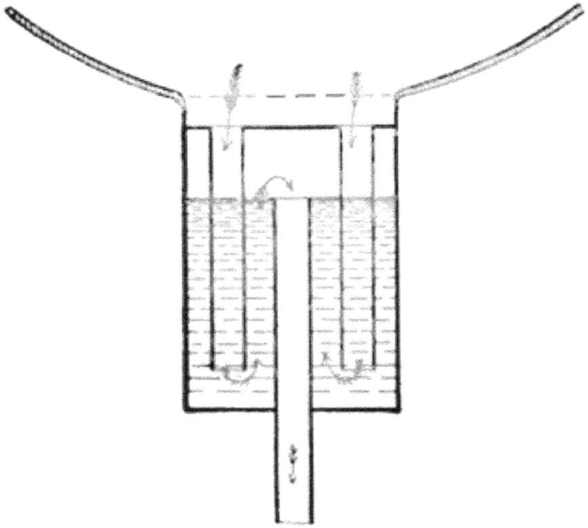

FIG. 159.—Renk's Trap.

rendered ineffective through evaporation of the the water. It is in use under sinks and urinals at the Hygienic Laboratory of the University of Munich, and was devised by its founder, the well-known Prof. Max von Pettenkofer, and in this place care is always taken to refill the trap attachment with water. For general use it cannot be recommended.

Dr. Renk, a pupil of Pettenkofer, has suggested the trap, Fig. 159, designed with the special object to be non-siphoning. I do not think this reservoir trap is self-cleansing, nor adapted to bowls or bath tubs.

CHAPTER VI.

INSECURITY OF THE COMMON WATER SEAL TRAPS.

OF all traps illustrated in the foregoing chapter, none is superior in point of cleanliness to the common S-trap. Experiments with such traps, however, have proven that they may become, in certain cases, most dangerous devices in a house, for the following reasons :

1. Traps may be forced by back-pressure.

2. They may lose all their water, when their fitting is emptied, by the momentum of the water rushing suddenly through the trap.

3. Traps may be completely siphoned, or at least their water level lowered below the dip in the trap, by a flow from another fitting on the same branch pipe.

4. Traps under fixtures may be siphoned by a sudden flow through the main soil pipe, to which these fixtures are connected by branch wastes.

5. If fixtures remain unused for any length of time, the water of the trap may evaporate so much as to destroy the seal.

6. With traps on dead ends of pipes, or with unventilated soil and waste pipes, there is danger that the water of the trap absorbs soil-pipe gases, giving them off on the house side of the trap. Even germs of disease, although not transmitted through water without motion, are said by scientific investigators to be liberated from the water if such is

violently agitated, as, for instance, with traps under fittings, when a discharge through such fittings occurs.

These statements are also more or less true of other traps, such as bell-traps, bottle-traps, D-traps, etc. It was on account of these objections, chiefly, that the more complicated *mechanical* traps were invented. On the other hand, the discovery was soon made that the danger from siphonage might be greatly lessened, under certain conditions, by emptying each branch waste pipe independently into the main soil or waste pipe. This, however, cannot always be done in buildings, nor is it, where it can be done, a protection against loss of water seal in all cases. The formation of a partial vacuum, and therefore siphonage, can, in most cases, be prevented by attaching a vent pipe *of suitable diameter* to the crown of the traps, leading its open end to the outer air (see Fig. 160). In the first place, such a vent pipe renders traps of any kind practically safe against siphonage, provided its size is such as not to offer too much frictional resistance to the air passing through it to break the suction. It is quite apparent that the diameter of the air pipe must be increased in proportion to its length, or, what means the same, to the height of the building. I lay particular stress upon this point, which does not seem to have been more generally understood.

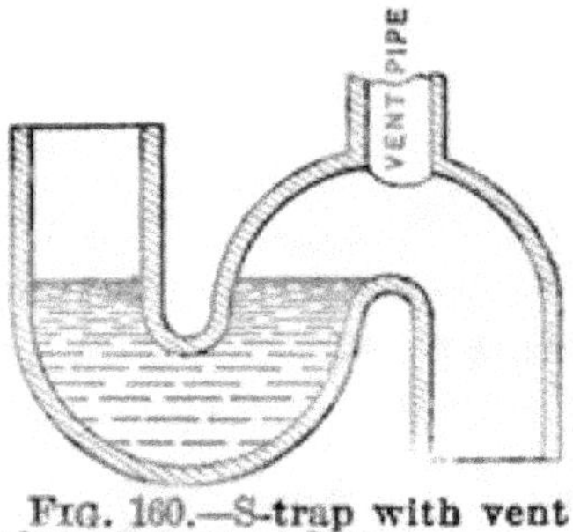

FIG. 160.—S-trap with vent pipe at crown of trap.

The vent pipe renders traps of any kind safe against back-pressure and absorption of gases.

Such air pipes, further, *prevent any stagnation of foul air in any part of the waste pipe system, and as such are powerful and important helps to a thorough ventilation of the drainage system.* This latter fact seems to have been overlooked by many who regard air pipes as safety attachments against siphonage *only*.

In applying such air pipes to traps, it is not necessary that each fixture should have an independent vent to the roof; several branch vent pipes may be connected into a main air pipe of ample size, and this may run along the soil pipe and may branch into it *above the highest fixture* (see Fig. 161), or else it may run above the roof independently.

While admitting that such air pipes render S traps practically safe against most of the above made objections, it cannot be denied on the other hand, that they largely increase the cost of plumbing in dwellings, especially so, where fixtures are much scattered throughout the house.

First, they complicate the work and are difficult to run in old buildings, and must be largely increased, in the case of high buildings, towards the upper floors.

Second, they increase the evaporation of water in traps, and therefore aggravate the danger from sewer gas entering through fixtures in cases where these remain unused for a long time.

Third, it is quite possible that vent pipes stop up in time at the crown of the trap with splashings from soap-suds, when they will cease to furnish air

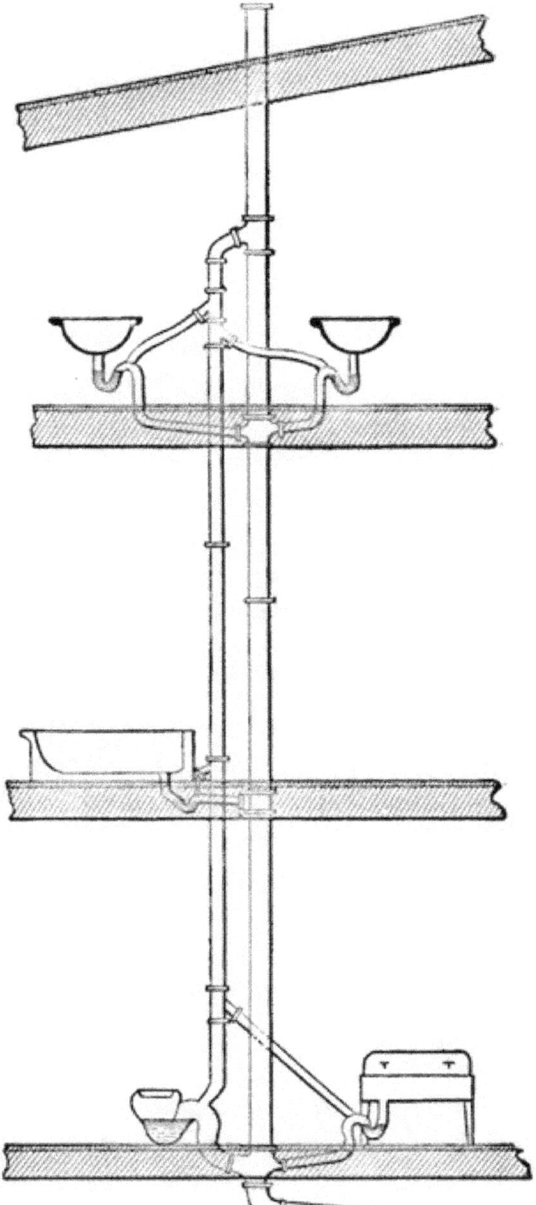

FIG. 161.—Stack of air pipe for a number of S-traps.

to break the vacuum. Unluckily, such fact would not reveal itself, and is not easily detected, nor is much known at the present time about this point.

The literature on this subject has been lately enriched by numerous, careful and valuable experiments upon the siphonage of traps, made by Col. Geo. E. Waring, Jr., assisted by the writer, by Messrs. Edw. S. Philbrick, C. E., and Ernest W. Bowditch, C. E., of Boston, by Mr. S. Hellyer of London, England, Dr. Lissauer of Dantzic, Germany, Dr. Renk of Munich, Germany, and others.

The results of the first mentioned experiments are greatly at variance, and seem to indicate, that while in some cases traps need a strong protection

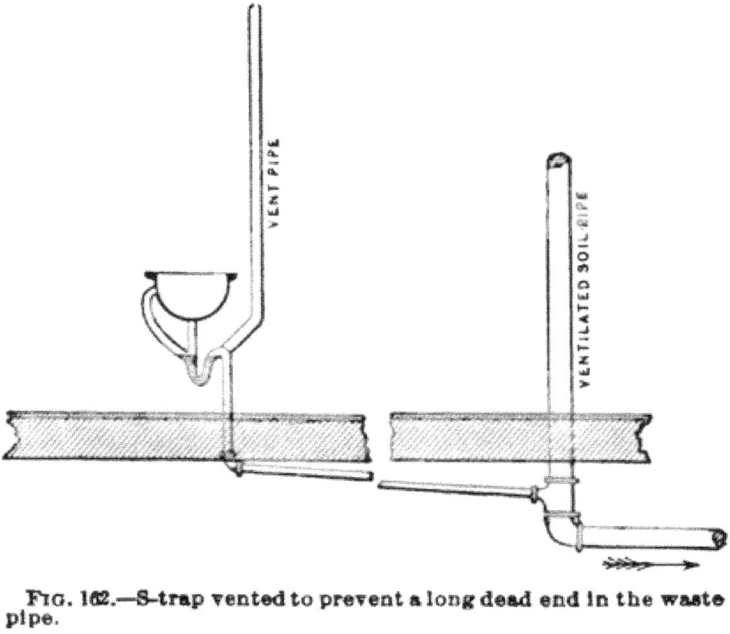

FIG. 162.—S-trap vented to prevent a long dead end in the waste pipe.

against siphonage, in other cases, especially where the soil and waste pipes have ample ventilation, and branch wastes are very short, such protection is not required. At any rate, it is too early yet to establish rules which apply to all cases. It has always seemed to me as if it would be feasible to practice a wise discrimination.

Where a fixture is located remote from a vertical pipe, and consequently discharges through a long run of waste pipe, which would otherwise form a "dead end" (see Fig. 162), it is *positively necessary* to run a vent pipe from the crown of the trap upward to the outer air, which prevents in the first place a stagnation of air, and at the same time stops siphonage; and this is true of *any kind of trap*, not only of the class of traps known as S-traps. It should apply to mechanical traps as well.

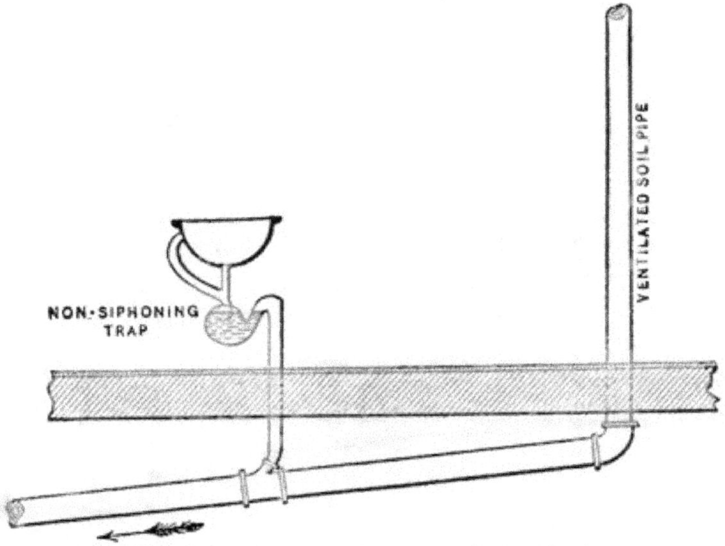

FIG. 163.—Non-siphoning trap under bowl, where this is near a thoroughly ventilated soil-pipe.

If, on the other hand, such fixture is located quite near to a vertical *thoroughly ventilated* soil pipe, or a well ventilated horizontal run of pipe (see Fig. 163), I should not hesitate to place under the fixture a trap which is not easily siphoned, leaving out the air pipe if there is no vent pipe near by to connect to. Such a course seems especially desirable in the case of high buildings for single fixtures in basements, or on lower floors. For instance, a 1½ inch sink trap in the basement of a flat, such as is now being erected in N. Y. City, 200 feet in height, would require an air pipe at least 3 or 4 inches in diameter to prevent siphonage, the friction in a 1½ or 2 inch pipe two hundred feet long being too great to allow the air to enter quickly enough to break the suction. I would consider it foolish extravagance to use such long length of pipe of such large size for the trap of only a single sink. If a non-siphoning trap could not be made to answer the purpose, the only sensible course to pursue would be to abandon such fixture entirely.

I must further say that it seems to me dangerous to use vented S-traps with the usual water seal of only 1½ or 2 inches under bowls or tubs in spare or guest rooms of large city residences, and for such dwellings generally that are occupied only a part of the year. This danger is generally disregarded or passed over lightly by enthusiasts for "back air piping." My personal preference in such cases would always be for a non-siphoning trap, with a water seal which does not so easily evaporate, or for a non-siphoning trap with a mechanical seal against gases from the soil pipe, and where rules of local Boards of Health would demand such an air pipe under such conditions, I should probably advise the use of a tight-shutting stop-valve on the waste pipe, and combined with it an arrangement for simultaneous shutting off the hot and cold water supply to the fixture, so as to render an overflow impossible. I am quite ready to admit that the latter arrangement would tend to complicate the plumbing work, but, I think, everybody must concede that, under the conditions mentioned, it would be safer than a vented S-trap with usual slight seal.

Mention has already been made in the foregoing chapter of *non-siphoning* traps. A bottle-trap, with a deep seal, will answer in many localities, and the Bower trap likewise answers for this purpose. Pietsch's trap may be called a non-siphoning trap, although objectionable on account of the flap-valve. There is Knight's trap, clumsy and not self-cleansing; Randolph's trap with double ball-valve; and Wm. Scarborough's trap. In Fig. 157 the writer has suggested a non-siphoning trap; some experiments, made with a rather primitive model, proved it to be efficient against siphonage. Morey's trap attachment (Fig. 155) may be efficient while new, but the valve may get out of order after some use, and then it would open a road to sewer-gas, without any warning whatever. Pettenkofer's trap attachment (Fig. 158) may answer very well where people will take the trouble to refill the trap attachment every other day, but for general use it is entirely unfit. Renk's non-siphoning trap (Fig. 159) is much to be preferred, but the objection must be made against it, that it becomes a reservoir for filth.

In this connection, I must make mention of an apparatus for preserving the water-seal of traps, called the "Eureka Trap Governor." This device is the invention of Mr. C. Lightbody of Brooklyn, N. Y. Its object is to prevent the loss of the water-seal in traps, either through evaporation or siphona e, by connecting the trap by a special pipe with the water supply, in such a manner that when any loss of water occurs, a new supply of fresh water is immediately admitted. Fig. 164 shows a

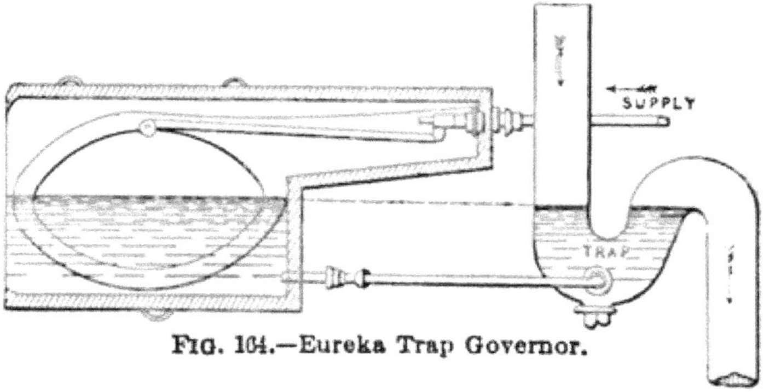

FIG. 164.—Eureka Trap Governor.

section and partial view of the apparatus. "A small cast iron tank (15 inches in extreme length by 7 inches in depth), made in two vertical sections of equal size, fastened together by screws, contains a copper float connected by a brass arm with a valve in a ¼ inch coupling to be connected with the water supply. From one end of the tank, near the bottom, a ¼ inch pipe is taken to the lower bend of the trap, as shown. The governor is so adjusted that any fall of water in the trap will open the valve until the entire seal is restored. The tank is to be fastened by screws on the wall near the trap, by means of lugs on the back side. In the front of the tank is a glass window showing a portion of the float; this is not seen in the drawing, as a larger surface is broken away to show the whole float." There can be no doubt about the efficiency of such a device, although it would somewhat increase the expense of fitting up plumbing fixtures. An objection to such an apparatus would seem to be the danger that the ball cock would leak, which fact would not become apparent. From a sanitary point, such a leakage would not be objectionable, as it would tend to change the water in the trap constantly, but this latter object could be just as well attained by keeping a dribbling stream running from the faucet into the fixture. Yet it must be said, that such arrangement would largely increase the waste of water, which is already a source of serious trouble to Water Departments; therefore, the apparatus is not likely to meet with favor.

Valuable scientific researches have also been made relating to the absorption of gases by the water in traps. Dr. Andrew Fergus of Glasgow was the first to experiment on this point, and his conclusions, though valuable, were modified and corrected by the results of experiments made by Dr. Neil Carmichael, Edward Frankland, Prof. Raphael Pumpelly, Dr. Wernich and Naegeli in Germany, and others.

It is now generally accepted that, with a thoroughly ventilated soil and waste pipe system, little, if any, absorption of gases by the water of traps occurs. Even should such water contain germs of disease, they are not supposed to be liberated from it unless the water is violently agitated. In other words, there is no sufficient reason for feeling anxiety in regard to absorption of deleterious gases by water-seal traps.

CHAPTER VII.

DEFECTS IN THE PLUMBING WORK OF DWELLINGS.

LEAD is the material usually employed for branch waste pipes connecting fixtures with the main soil pipe system. Lead pipe of small diameter is more easily run than an iron pipe, and although it is quite feasible to run asphalted wrought-iron waste pipes of small size to wash-basins, tubs, or sinks, it must be conceded that lead offers certain advantages, especially in crooked runs, in corners, and under floors. Foremost among the advantages should be mentioned the fact that the least number of joints are required with lead pipe. For more than one reason, however, it is very desirable to avoid supply and waste pipes in concealed places: it is a matter of common occurrence with lead waste pipes located under the floor, to have nails driven by a carpenter's careless hand, into the upper part of the waste. Unfortunately, such a fact is not generally discovered at once; the hole being on the top, it may not leak water, but it will certainly leak sewer-gas.

Where lead waste pipes escape such a treatment from carpenters, they are subject to the danger of being gnawed by rats. If concealed under floors, waste pipes are often run at a dead level, or where proper fall has been given to them at the time the work was done, a subsequent sagging may occur, owing to insufficient support, and the pipe, consequently, becomes double trapped or air bound. Here, as in regard to plumbing fixtures, the rule should be observed, to leave as much as possible in plain sight and open to inspection.

Defective joints in lead pipe are due to ignorance and inability of mechanics. Lead pipe should always be connected with "wiped joints," which technical expression means that the joint should be made with solder wiped in a shapely oval lump around it. Very often the back part of such joints is found defective, the solder having dropped off. Where joints are out of sight, the wiped joint is usually carelessly made and unevenly shaped; but oftener still the plumber rests satisfied with making a "cup joint," which is not as strong nor workmanlike in appearance as the wiped joint.

Where lead pipes are joined to hubs of cast-iron pipe a careless workman often inserts the lead pipe into the iron hub, filling the space with cement or putty. Such joints are not to be trusted, as putty and cement crumble away in a short time, thus allowing the escape of noxious gases.

The proper way to make such joints is to use a tinned brass ferrule, which is inserted into the cast-iron hub, the joint being thoroughly caulked as in iron pipes; the lead pipe is connected to the brass ferrule by a wiped joint. Where lead pipe joins a wrought-iron or brass pipe, the connection is made with a brass screw nipple, soldered to the lead and tightly screwed with red lead into the iron or brass fitting, which is tapped to the standard thread.

A radical defect exhibited in the common systems of plumbing is the use of waste pipes much too large for the office they have to perform. Think of a 2 in. lead waste pipe for a single wash-bowl having only a $1\frac{1}{4}$ in. coupling and strainer, or a 3 in. waste for laundry tubs! Such pipes cannot possibly remain well flushed, but must soon become coated with filth, or even clog up entirely.

For ordinary pressure of water in the supply mains a $1\frac{1}{4}$ in. waste for a bowl is ample; a $1\frac{1}{2}$ in. pipe empties a bath-tub or a laundry tray as quickly as any one may desire; even for a pantry or a kitchen sink anything beyond $1\frac{1}{2}$ in. is a positive injury, and larger wastes are sure to choke up with grease in a short time. Not only is the first cost of the lead piping greater, but such extravagant sizes lead to stoppages and consequent bills for repairs, which can be avoided by doing the work right in the first place. But even in these enlightened days it is rare to find house-owners who will listen to disinterested advice on such subjects. Most of them still prefer to pay the price of the larger pipe in order to be sure that their waste pipe is "big enough to pass anything coming into it."

Another mistake frequently made is to use for such waste pipes traps of a larger diameter than the pipe. A short time ago I had occasion to examine the plumbing in a country residence that had just been completed for a wealthy New York merchant. The lead waste pipe from the kitchen sink was 2 in. in diameter, with a 3 in. trap; the bowls had $1\frac{1}{2}$ in. wastes and 2 in. Du Bois traps; the bath-tub had 2 in. wastes and a 6 inch bottle trap; the waste from the laundry tubs was 3 in.

in size with a 4 in. trap. Under no circumstance whatever should any trap be of larger calibre than the waste pipe; in my own practice I prefer to reduce the size of the trap $\frac{1}{4}$ or $\frac{1}{2}$ in., in order to increase the scouring effect of the waste water.

It would be impossible completely to enumerate all defects found in the plumbing of city and country dwellings. Some of the graver and more common faults have been explained in the foregoing chapters. Of others I merely mention: the connection of drip pipes to the soil pipe system or to traps; the connection of the overflow from drinking water tanks or water-closet cisterns to soil or waste pipes; the direct connection of refrigerator wastes to any part of the drainage system; the running of vent pipes from closet bowls into soil pipes; the running of soil or waste pipes into chimney-flues; the use of rain leaders as soil and waste pipes; the use of soil pipes as rain leaders; the use of rain leaders as only ventilators of house drains; untrapped leaders opening near dormer windows; the trapping of fixtures at a distance from the soil pipe; the use of one trap for a number of fixtures; the double trapping of fixtures; the running of air pipes for traps into ventilation flues; the connection between air pipes from traps and vent pipes from closet bowls; the junction between air pipes and traps made on the wrong side of the trap, etc.

With such a large and by no means exhausted list of possible defects in the plumbing of a house, the importance of a general house ventilation cannot be too often stated. The occupants of a house may sometimes continue to enjoy good health in the face of such dangerous defects as long as ample provision is made for ventilation, and so long as a current of pure air daily sweeps through all rooms and closets of a dwelling. With no exit for foul air, let the poison once accumulate in a house and the consequences may be serious.

The conclusions which may be drawn from what has been said above are two-fold, namely : first, that by providing a dwelling with modern conveniences, having for their object comfort, cleanliness and promotion of health at home, we also create the danger of air pollution in dwellings, and that although it is quite possible to have such fixtures well and safely arranged, such a result can hardly be expected from the average mechanic, and that the best course for a house-owner is to procure professional advice at an early stage of house building. Second, that no matter how well the system may have been planned, conceived and constructed, it needs looking after from time to time, same as any other engineering structure, and just here let me remind the reader of the importance of having on permanent record the location of all pipes, fixtures, traps, etc., inside a dwelling, in order to facilitate inspection and repairs.

It may not be inappropriate to close this chapter with the following excellent remarks from Dr. Simon :

"A very large danger to the public health, and particularly to the better off classes of society, has of late years consisted in the recklessness with which house drains, receiving pipes from water closets, sinks, cisterns, baths, etc., in the interior of houses, and often actually within bed-rooms or the adjoining dressing-rooms, have been brought into communication with sewers. *Among architects and builders there seems to have been very imperfect recognition of the danger which this arrangement must involve, in event either of unskillful first construction or of subsequent mismanagement or want of repairs.*

Then, in regard to construction, an almost unlimited trust has been placed in artisans who not only could hardly be expected to understand certain of the finer conditions (as to atmospheric pressure) which they had to meet, but who also, in not a few instances, have evidently failed to apprehend that even their mechanical work required conscientious execution. (The italics are mine). Under influence of the latter deficiency, there have been left in innumerable cases all sorts of escape holes for sewer effluvia into houses, and disjointed drains effusing their filth into basements: while, under the other deficiency, house drainage, though done with good workmanlike intention, has often, for want of skilled guidance, been left entirely without exterior ventilation, and sometimes has, in addition, had the overflow pipes of baths or cisterns acting as sewer ventilators into the house; and all this not infrequently in places where the sewer itself, from which so much air has been invited, has been an ill-conditioned and unventilated sort of cesspool.

It is almost superfluous to say that under circumstances of this sort, a large quantity of enteric fever has been insured, and I should suppose that also a very large quantity of other filth diseases must have sprung from the same cause. Then there has been the vast quantity of interior air fouling which arises from mismanagement of drain inlets, or from non repair of worn out apparatus; as when sink traps, injudiciously made movable, have been set aside; or when pipes under temporary disuse, having evaporated all water from their traps or leaden closet pipes, with holes corroded in them, have been left fouling the house with a continuous eructation of sewer air. Again, in poor neighborhoods, water-closets have, in many cases, been constructed with scanty and ill-arranged water service to flush them or have even been left to only such flushing as the slop water of the house or the other water thrown in by hand might give; and again and again these ill-watered and often obstructed closets have been found acting, on a large scale, as causes of disease.

Again, a different sort of danger, and one which seems capable of wide operation, has been seen to arise where water-closets received their water service from the mains of a so-called 'constant' supply, for supplies called constant must not only sometimes intermit for purposes of necessary repair, but also in some cases are habitually cut off during the hours of night, and the danger is that during times of intermission, if there be not service boxes or cisterns between the privy taps and the main privy, effluvia, and even in some cases fluid filth, will be (so to speak) sucked from closet pans into water pipes."

CHAPTER VIII.

CELLAR DRAINS AND DRAINAGE OF CELLARS.

BEFORE proceeding to examine the external sewerage of houses, let us descend into the cellar, for, although commonly it is the most neglected and least thought of part of a dwelling, its sanitary condition has a direct bearing upon the well-being of the occupants of the house. I think I am not mistaken in saying that from the condition of a cellar one may, with tolerable accuracy, draw conclusions in regard to the healthiness of the whole house. In the first place a cellar should be thoroughly ventilated, for much of the air of a cellar is drawn into the upper rooms of a house, particularly in winter time, when stoves and fireplaces create a constant suction toward the rooms.

Moreover, where hot-air furnaces are placed in the cellar, these generally draw their air supply directly from the cellar, or, where a cold air box has been provided, it is constructed of wood, in a wretched manner, being full of cracks and open seams, through which the tainted atmosphere of the cellar enters, to be carried in a heated state to the upper floors of the dwelling.

It is all-important that the cellar floor should be thoroughly dry and tight; nothing is more injurious to health than ground-air, which is often tainted with sewer gases from leakage of drains, or from cesspools, located under the cellar of a house, or from heaps of garbage and refuse, constituting the soil upon which many of our habitations are being constantly erected, notwithstanding all earnest protests from the most prominent sanitarians. A cellar should be well-lighted, for this will aid in keeping it in good order and will promote cleanliness. Cellars should never be used for the storage of vegetables, nor should any kind of rubbish be left there to decompose. They should not be made hiding places for old rags, worn-out clothing, tin-cans; and, above all, the darkest corner, or the place under the cellar stairway, should never be chosen for a servants' water closet.

If the cellar is low, and apt to be damp or even wet at times, proper drainage must not be neglected. But under no circumstances whatever establish a *direct* connection between the cellar, or the subsoil under the cellar, and the sewer. You invite sewer gas into your house by doing so. Do not place any reliance upon the common S-trap, with shallow water seal, on the line of the cellar drain; it is too often rendered useless by the evaporation of the water forming the seal. Still worse is the common so called "cesspool or stench trap"

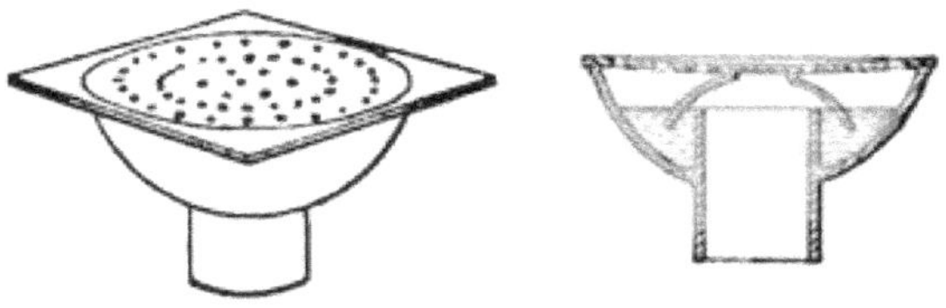

FIGS. 165 and 166.—Cesspool or Stench Trap.

for cellar floors, provided with a bell-trap

of improper shape and much too insufficient water seal, which is often rendered ineffective when the loose strainer gets lost. If there must be an opening in the cellar floor to remove water after scrubbing floors, or in case of an unexpected leakage of water into the cellar, this opening should be of moderate size and covered with a strainer, and the branch drain leading from it to the main house drain should be trapped by a trap with very deep seal, not liable to be easily lost through evaporation. The author uses in his own practice an S-trap for cellar floor drains, which has a depth of water seal of six inches, and presents a small surface, so that it is not easily affected by evaporation. Should, however, the sewer in the street be

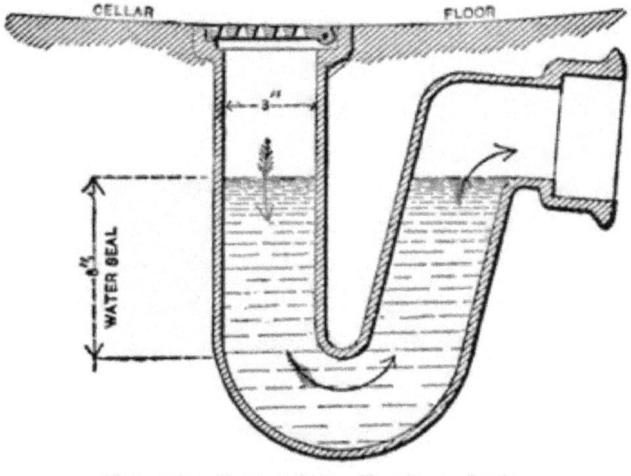

FIG. 167.—S-trap for cellar floor drains.

subject to back-flooding from the tide or an unusual rise of a river, or should its size be insufficient to carry off heavy rainstorms, the cellar would be in constant danger of being flooded by backwater and sewage, in which case—upon the water receding—deposits of foul matters are left on the cellar floor. In such cases I strongly advise doing away with the opening in the cellar floor, or else I should insist on the use of some back pressure or tidal valve on the drain outlet.

The cellar is usually the place where the various soil and waste pipes of a dwelling are connected or combined into one main drain, the cellar or house drain.

Mr. Dempsey, an English civil engineer, speaks about the arrangement of house drains as follows, in his book, "Drainage of Towns and Buildings:"

"The first step in the arrangement is to collect the whole of the drainage to one point, the head of the intended draining apparatus, and the determination of this point requires a due consideration of its relation to the other extremity of the drain at which the discharge into the sewer is to take place. In buildings of great extent this will sometimes involve a good deal of arrangement, and it will, perhaps, become desirable to divide the entire drainage into two or more points of delivery, and conduct it in so many separate drains to the receiving sewer. The length of each drain being thus reduced to a manageable extent, the necessary fall will be more readily commanded, and the efficiency of the system secured. * * If the rain water falling on the roof of the building, and on the yard or space attached to the house, is not applied to any other purpose, it will have to be conducted into the drain to be discharged with the sewage. These waters, being purest of the contents, should be received as near as possible to the head of the drain, and made to traverse its entire length, and thus exert all the cleansing action of which they are capable."

In most houses built more than five years ago

you will find the main drain buried below the floor, in inaccessible locations, its position being often quite unknown. In such houses glazed earthen or cement pipes are used for house drains, but the drain of old buildings was usually built of brick, often square in shape (see Fig. 168), much too large in size, and with insufficient or no fall. Sometimes troughs of wood were used to carry off waste waters. All such drains are sure to accumulate deposits and to generate disease-breeding gases of decay.

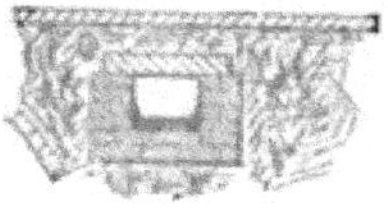

FIG. 168.—Brick Drain.

Brick drains under houses are generally harboring places for rats, the cement of the joints crumbles away, bricks loosen and fall out, and the drains become leaky, or partly choked. Sometimes, in examining old houses, I have observed that vitrified pipes had been laid under the floor to take the place of the brick drain, the latter being simply cut off, but left, full of decomposing filth, under the building. Even vitrified pipes of proper shape should never be used for drains under a dwelling house; they often crack through settlement, and have leaky joints, and the floor under cellars becomes saturated with sewage. It is impossible properly to connect an upright soil or waste pipe to an earthen drain, for no matter how well the iron pipe may be cemented into the hub of the terra cotta drain, a settlement of the soil pipe will break off the hub; in other cases the earthen drain settles away from the soil pipe, leaving an opening between both, through which all sewage matter is discharged onto the ground under

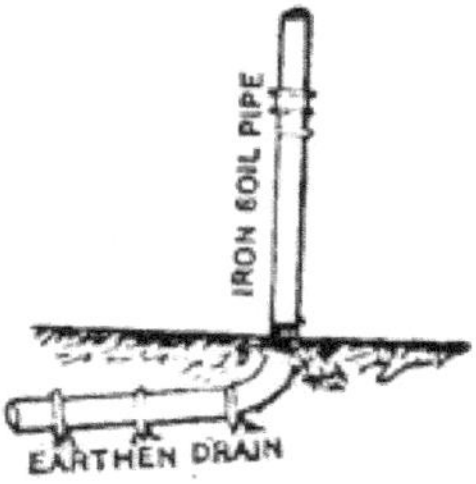

FIG. 169.—Earthen Drain.

the cellar. The author lately had occasion to see such a defective connection in a fine residence on Madison avenue (see Fig. 169).

If the house drain must be laid under the cellar floor it should consist of heavy iron pipes with well-tightened joints and should be made accessible in a few proper places to provide means for removing accidental or malicious obstructions or stoppages, which are liable to occur even in the best devised and best constructed system of house drainage.

The necessity of running the house drain below the cellar floor exists only in rare cases. In most cases it is possible to banish plumbing fixtures from cellars, to find a better lighted place for the laundry and washing tubs, and a place for the servants' closet free from the objections heretofore made. In such case it is best to run the house drain of iron pipe across the cellar, either along a foundation wall, or suspended from the ceiling. This brings the drain in sight for inspections, and it is a recognized principle of modern house drainage that as little as possible of waste pipes and of the plumbing work in general should be hidden from view.

CHAPTER IX.

USUAL DEFECTS OF HOUSE DRAINS; SEWER CONNECTIONS; PRIVY VAULTS AND CESSPOOLS.

IT behooves us now to inquire into the *external* sewerage of the dwelling. Faults of the interior drainage work contribute, as we have seen, a large share to the pollution of the atmosphere which we breathe; faulty external sewerage, besides being the cause of a vitiation of the air, creates a most dangerous pollution of the soil around and under habitations, and likewise frequently poisons the water from wells and springs.

Hence it is a mistake, which, however, is frequently made, especially in rural districts, to neglect the outside drainage of a dwelling. The water which we drink must be as pure and wholesome as the air we breathe; and since country houses depend most always upon a well or cistern situated near the house for the supply of this indispensable element, the external sewerage of such houses is of greater importance even than that of city houses. But in both cases the teachings of sanitary science require a proper care in laying such external drains which should remove at once from habitations all sewage matters.

The defects usually found in external drain pipes are numerous. They relate to the construction of the drain, to the manner of jointing pipes and laying drains, to the materials used for such drains, to their size and shape, and to junctions with branch drains and with the street sewer. Foremost among them I mention *leaky joints*, for these work multifold harm. Not only does the liquid soak away into the soil to find its way to the nearest well or spring, but a constant accumulation and gradual saturation of the soil with filth is inevitable. Again, deposits will occur in the pipes, as the force of the flush is to a great extent lost, if the waste water soaks away at the joints, and the solid part of the remaining sewage in the pipes must soon decompose and fill the pipes with gases of decay.

A second cause of deposits in house drains is an irregular or insufficient inclination of the pipe. How seldom it is that the simple precaution is observed of taking levels to ascertain the available fall from the point where the drain leaves the house to the junction with the sewer. Hence we often find house drains sloping the wrong way, being in reality nothing but "elongated cesspools." How easy would it be to avoid such mistakes by the use of even a common spirit level!

A further grave defect is the almost universal preference of drain-layers and ignorant builders for *large pipes*. Not many years ago nine and even twelve-inch pipes were used for the drainage of an ordinary city house and lot; only lately six-inch house drains have been used for the average sized city dwelling, and a four-inch pipe, which answers for most city or country houses, except for unusually large residences, is still the exception. The larger the pipe for a given amount of water the more sluggish will the velocity of the stream be, and thus we find in the large size of drains another cause of accumulation of deposits.

Mr. Dempsey, C. E., in his "Drainage of Towns and Buildings," says:

"Sewers and drains were formerly devised with the single object of making them *large enough*, by which it was supposed that their full efficiency was secured. But sluggishness of action is now recognized as the certain consequence of excess of surface, equally as of deficiency of declination. A small stream of liquid matter, extended over a wide surface, and reduced in depth in proportion to the width, suffers retardation from this circumstance, as well as from want of declivity in the current. Hence a drain which is disproportionally large in comparison to the amount of drainage, becomes an inoperative apparatus, by reason of its undue dimensions, while, if the same amount of drainage is concentrated within a more limited channel, a greater rapidity is produced, *and every addition to the contents of the drain aids, by the full force of its gravity, in propelling the entire quantity forward to the point of discharge.*" (This latter point is especially little understood).

The English architect Ernest Turner, well known as a prominent sanitarian, speaks about size of drain pipes as follows in his book, "Hints to Househunters and Householders":

"It is extraordinary that the practice of making drains as large as possible instead of as small as may be necessary for efficient working should have continued so long as it has. The only possible reason must be, 'every drain is bound to choke sooner or later, and the larger the pipe the longer it will take before it requires cleaning.'

"The smaller the pipe the less the friction—*the greater the hydraulic pressure the greater the velocity*, and consequently the less chance there is of any obstruction taking place.

"It is a common notion that an ordinary medium-sized dwelling-house requires a nine-inch drain; but the idea is altogether erroneous.

"To carry off a small quantity of water quickly, a small pipe must be used. The greater the proportion of the wetted perimeter to the volume of water to be discharged, the greater, obviously, the resistance.

"If a pipe becomes choked, it is generally owing to its being too large—not too small—or to faulty laying or construction."

Mr. Eassie, in his chapter on House Drainage, written for the recently published book, "Our Homes, and How to Make them Healthy," has the following:

"Drains are very frequently laid down of far too large a sectional area: six inches in diameter where four inches would have sufficed, nine inches where six inches would have been sufficient, and twelve inches where nine would have been ample. This laying down of too large pipes is one of the most besetting sins in house drainage, when that has been left entirely in the hands of the builder. I have taken up twelve-inch pipes in a house, and replaced them with six-inch pipes. The sizes of the pipes to be used should not be decided hap-hazard, but advice taken upon this subject from a competent person. As a general rule, a four-inch pipe is sufficient for a cottage, and a six-inch pipe for an extensive dwelling. In deciding the diameter of the drain pipes, due provision must be made for the rainfall, or serious floodings may be the result after every storm of unusual severity."

Other defects of house drains relate to the shape and material of the pipes. Brick drains with flat bottoms are an abomination, but some of the finest residences of Fifth avenue remove (or rather retain!) the household wastes through such square channels, 12″x12″ in cross section. Wooden drains are not any better. Being alternately wet and dry they quickly rot; the roughness of the inner surfaces of such conduits tends to create deposits.

Vitrified pipes, properly shaped, smoothly glazed and well-burnt, are preferable to all other kinds, even to cement pipes. They should, however, be laid with care, on proper foundations, properly supported, well aligned, properly jointed and laid with a regular fall. Often no attempt is made in tightening the joints of vitrified pipe, and the mistaken notion largely prevails that through such open joints subsoil water may be removed, the house drain thus performing a double service, for which it should never have been intended. Conduits for the removal of the foul liquid wastes from houses should be *tight* beyond doubt. In made ground, where drains are liable to settle and break, earthen pipes should not be tolerated, but must be replaced by iron pipes, and this is true as well for drains passing near a well or cistern. And here the same remarks heretofore made as regards the quality of iron pipes might, with advantage, be repeated. Radical and thorough improvements in the make of iron drain pipes are much to be desired.

Another serious and frequent defect relates to the junction of branches to the main drain, T branches, *i. e.*, right-angled connection pieces being used, which cause eddies and accumulations of deposit. The same error of construction is often made at the point where a house drain connects to a street sewer. In order to join the flow from both with the least possible retardation of the current, the branch should enter the drain under an angle of 45° or 60°. (See Fig. 170.) I must not forget to mention an additional defect, namely, that of delivering a large drain pipe into one of smaller diameter, a mistake too often made by ignorant or skin builders.

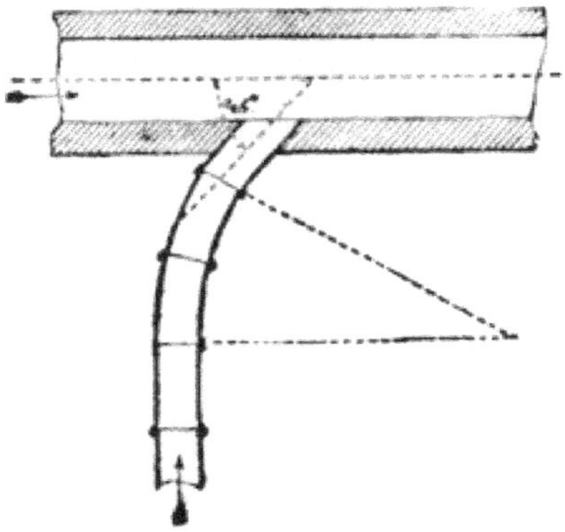

Fig. 170.—Proper manner of connecting house-drains to sewers.

It is interesting, though somewhat sad to learn that the defects in house drainage just described have not by any means been recognized only lately; for as early as 1852 the General Board of Health of England discussed the question of house drainage in an able and thorough report, arriving at exactly the same conclusions and principles which the best modern talent advises. I give below a few extracts from the Report :

"The materials of which house drains are commonly constructed are burnt clay bricks, and of these bricks for the great majority of houses, any inferior rubbish that can be put away, is used. The common 'place brick' is so absorbent and permeable that each brick will usually absorb about a pint of water. It is rough and ill-formed on the surface, so as to impede the flow of the sewage. The bottoms of the drains of houses occupied by the poorer classes are not always formed of whole bricks, brick-bats being often used for the purpose, which are frequently put together dry, or the mortar used for their connection is inferior, soluble and permeable to water as well as to gases. The following (Fig. 171) are common forms of permeable brick drains, which let out offensive liquid to spread beneath the premises, but detain, like sieves, the solids or less soluble matter:

FIG. 171.—House drain of brick, square in shape.

In many of the large provincial towns visited, still inferior drains are constructed. The following (Figs. 172 and 173) are two specimens:

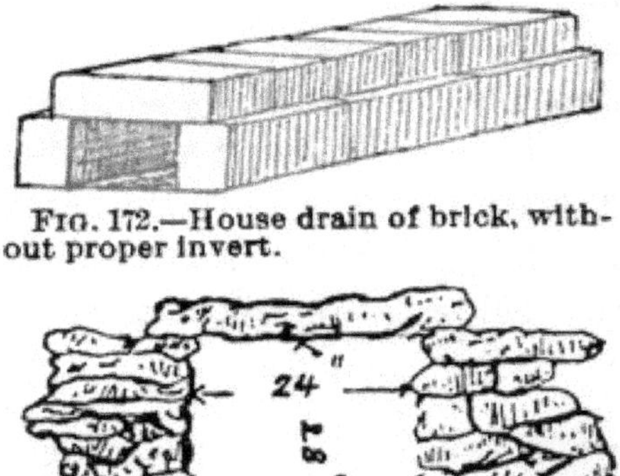

FIG. 172.—House drain of brick, without proper invert.

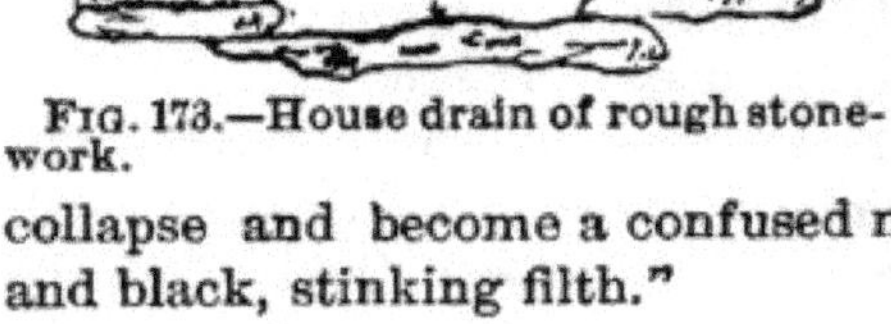

FIG. 173.—House drain of rough stonework.

The former will often be choked up in a few months, especially if some other owner, as is often the case, drains into it. The latter, it must be obvious, will ultimately have the same fate, notwithstanding the supposed advantage of its large size—should it not sooner collapse and become a confused mass of rubble stone, and black, stinking filth."

In regard to extravagant sizes for sewers and drains, the report says:

"It is important that the result of inquiry on this point should be understood,—namely, why a small channel or drain, properly adjusted to the run of water to be discharged, will be kept clear, while a large channel, with the same quantity of water to be discharged, and with the same fall or inclination, will accumulate deposit.

In large drains a given run of water is spread in a thin sheet, which is shallow in proportion as the bottom of the drain is wide, hence friction is increased, the rate of flow retarded, and according to a natural law, matters at first held in suspension, and which a quicker stream would have carried forward, are deposited. If there be any elevated substance, the shallow and slow stream, having less velocity and power of floating or propelling a solid body, passes by it. Thus, if by any neglect substances not intended to be received by a drain enter it, for instance, if a scrubbing-brush or hearth-stone has been allowed to get into, say a 15-inch drain, the height of water in regard to such substance may be as in the following sketch, fig. 174:

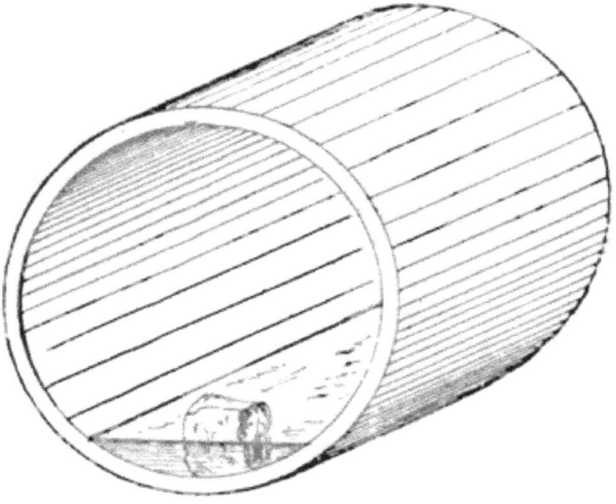

FIG. 174—A 15-inch house drain, with a shallow stream of water.

But if it were a 4-inch drain, the same quantity of water would assume a very different relative position, as in this smaller sketch, Fig. 175, and it will be readily

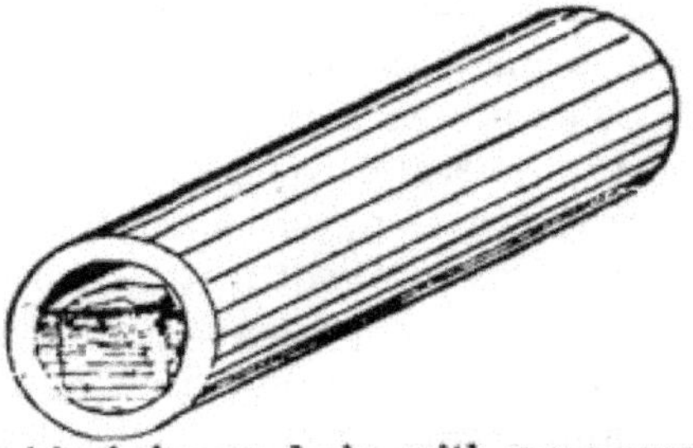

FIG. 175.—A 4-inch house drain, with same amount of water running through it as passes through the 15-inch pipe.

understood that the deeper stream of the contracted channel would be more powerful to remove any obstructing body.

Instead of concentrating the flow of small streams, and economizing their force, the common practice is to spread them over uneven surfaces, which "deadens" and "kills" them.

In a small drain an obstruction raises an accumulation of water immediately, which increases, according to the size of the obstruction, until four, five or six times more hydraulic pressure is brought to bear for its removal than could by any possibility be the case in a large drain; for in a large drain of three or four times the

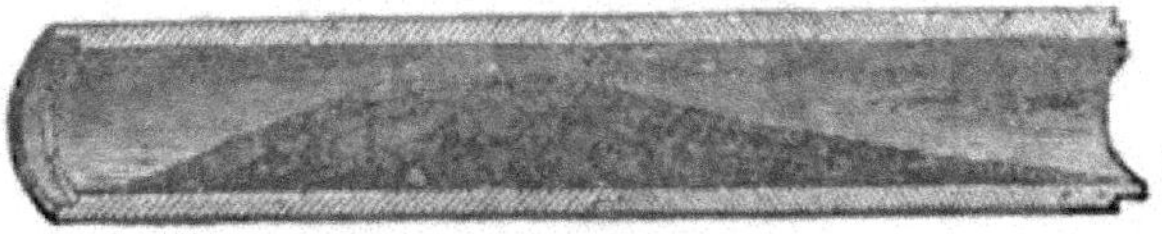

FIG. 176.—Accumulation of deposit in a house drain.

same internal capacity, the water can only be dammed up to the same relative height by an accumulation of matter three or four times higher, and therefore 27 or 64 times greater, which will gradually lengthen out, as shown in sketch, Fig. 176, and then be beyond the power of removal by the water."

From personal notes of a recent inspection of the drainage of a large sea-side hotel on the Atlantic coast I quote the following:

"The ground underneath the buildings appears to be saturated with excremental and greasy filth. There is an extensive network of terra cotta drain pipes under them, a few of these being main lines, into which a large number of laterals discharge. Most of these laterals are six inches in diameter (sometimes for a single kitchen sink), but some are even larger. These drains are laid in the most wretched manner, without regard to alignment or grade, partly on the surface, partly in the ground, a few being only half-covered. Few, if any, joints appeared to be tight. I observed the rising tide coming out of some joints in a heavy stream; in other joints the cement had crumbled off, or had been washed out or was removed through gnawing of rats. Many drains were cracked and broken, some had large holes at the top, which allowed sewer air to pass freely upwards into the buildings. Laterals joined the main sewer pipe by T-branches. I could not detect a single Y-branch; some laterals even run into the main drain in a direction *against* the current. The whole drainage work under the building appeared to be patch-work, done from time to time as occasion required. The waste-pipes from fixtures located in the building delivered directly into the network of drains just described; all kinds of materials were used for such wastes: square wooden pipes, galvanized iron pipes, tin, lead, cast and wrought iron and earthen pipes. A bend at the junction of a vertical and a horizontal pipe was the exception; most junctions were made with right-angled elbows. The only ruling principle for the drainage of the building seemed to have been to provide drains of *ample* size. At times of high tide the sewage backs up in the drains and floods the surface under the building oozing out at most of the joints. When the tide recedes, sewage mud is left on the ground to decompose. Hence arose the frequent complaint of offensive smells from the drains."

Conditions such as are described in these notes

are by no means exceptional, and similar defects exist in most houses at the present day. Owing to the indifference of the general public the actual condition of the drainage of a house is something seldom inquired into, except when sickness has made its appearance, or continued complaints of ill-health force it to the attention of the house occupants.

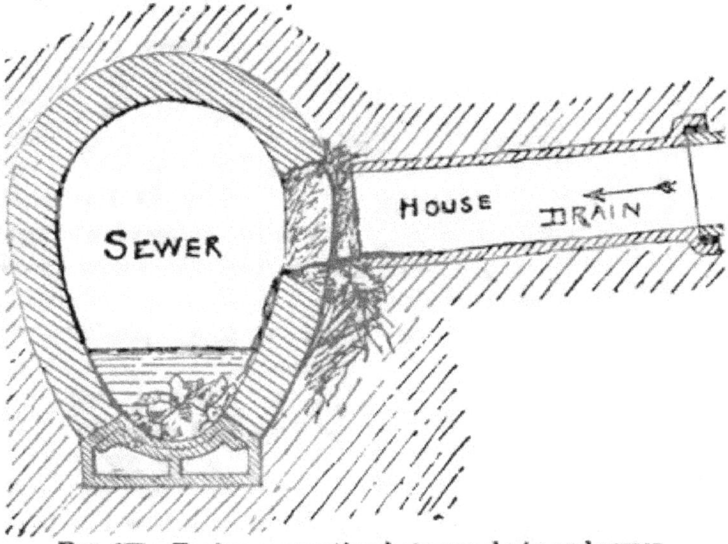

FIG. 177.—Faulty connection between drain and sewer.

And now I must offer a few closing remarks about the usual modes of disposing of liquid household wastes and human excreta. Comparatively few cities have as yet constructed a complete sewerage system with sewers in all principal streets, to which the house-drains connect. Many cities, however, are provided with a partial system of sewers, more or less faulty in design and worse in construction. With these it is a common occurrence to find the connection between house-drain and sewer improperly made; the following sketches, Figs. 177, 178 and 178a, which I borrow partly from Hoskins' "An Hour with a Sewer Rat," partly from Eliot C. Clarke's "Common Defects in House-drains," illustrate such faulty connections.

Many cities remain, up to this time, without any system of sewerage whatever, and in smaller towns and villages it is a common, though much to be condemned, practice, to store the sewage of the household in cesspools, which are not unusually located close to the house, in some cases even

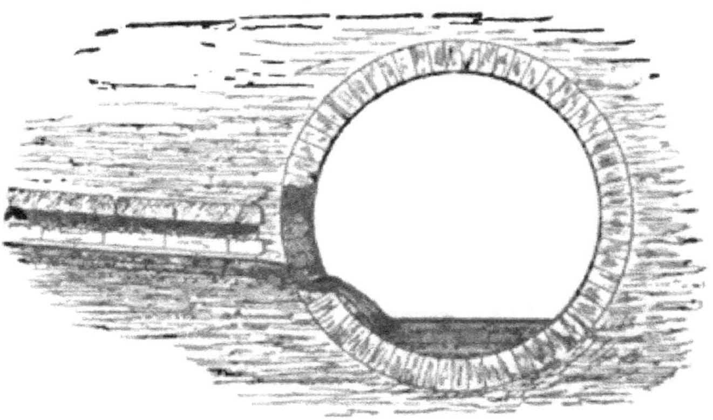

FIG. 178.—Faulty connection between drain and sewer.

underneath the dwelling. In most cases cesspools are mere pits, dug in the ground and walled up with loose stones. The liquid contents are left to soak away into the subsoil, while all solids and grease from the kitchen remain in the cesspool to decompose and generate noxious gases. Should the pores of the soil stop up and the liquid cease to leach into the ground, the cesspool is abandoned, generally, and a new hole dug, close to the first one. In other instances two cesspools are built, the first one, supposed to be tight, to retain the solid and grease from the household, the second one a leaching cesspool, connected with the first one by an overflow pipe, through which the filthy liquids run to be disposed of by soakage into the ground. A continuous pollution and dangerous saturation of the soil about human habitations is thus going on, while the air which we breathe is tainted by the foul emanations commonly knows as "sewer gas."

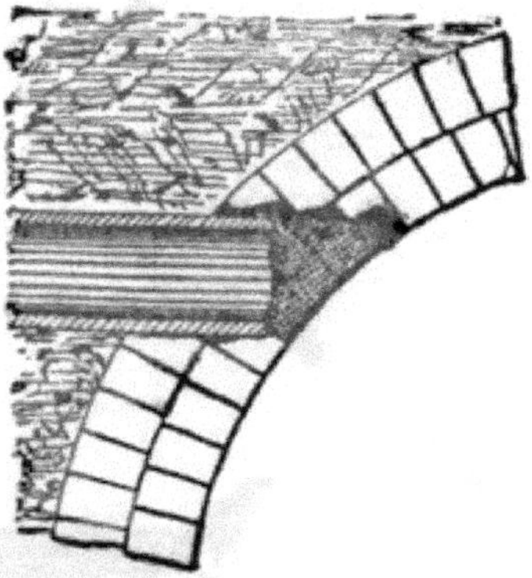

FIG. 178A.—Faulty connection between house drain and street sewer.

The Report of the General Board of Health of England referred to above, says :

"House-drains, constructed as described, commonly convey the sewage into cesspools, from some of which the overflow is carried away into the sewer; but often there is no overflow drain and the liquid percolates into the soil beneath and adjoining the building. When the cesspool becomes filled with the solid filth detained, it is not unusual, instead of emptying it, to form another. Beneath many of the more moderate-sized houses as many as three cesspools have been found; their ordinary state is displayed in the following sketch, Fig. 179."

Not less dangerous than the accumulation of

putrid organic matter is the pollution of the underground water by the filthy liquid soaking into the ground. Chemical analysis of the water of wells, situated in proximity to cesspools, or receiving the surface drainage from stables, cow-houses, etc., most always reveals organic matter in the water. Such contamination is all the more serious, as in

FIG. 179.—House drain delivering into a filled cesspool.

towns and villages or isolated country houses, people quite often must depend upon the well for the supply of drinking water to men and animals.

Another much to be detested practice, which might almost be called a crime, is the use of an abandoned deep well for a cesspool. And this is true for drains discharging water closet wastes as well as those discharging slopwater only. Practically, there is hardly any perceptible difference between either kind of wastes after having been retained for some time in a cesspool.

The question is often asked: "*At what distance from a well would it be safe to put a leaching cesspool?* Sanitary science has but one answer to this query: *it prohibits the use of leaching cesspools altogether.*

Prof. Kedzie of Michigan has lately illustrated the question of soil and water pollution by showing two cones, one of which he calls the *cone of filtration*, Fig. 180, and the other the *cone of pollution*, Fig. 181. The first cone shows the distance and the area drained by a well. It is clear that the radius of the base of the cone must depend on the depth of the well, and on the character of the soil through which the well is sunk. Again, the area of soil which may be polluted by soakage from a cesspool will depend on the soil, and on the depth of the cesspool.

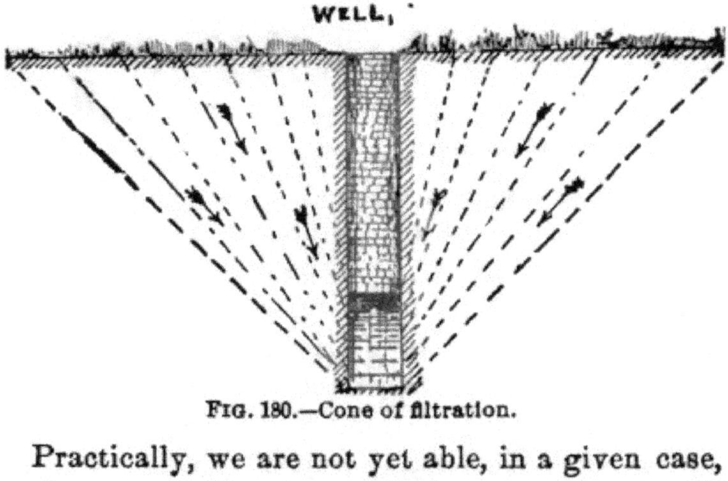

FIG. 180.—Cone of filtration.

Practically, we are not yet able, in a given case, to draw exact diagrams of both cones; if we could do so, the question of pollution of a well, in a given instance, by soakage from cesspools, could easily be answered without the aid of chemical analysis. The diagrams, however, are admirably adapted to convey to the general public an idea of the danger incurred by locating wells and cesspools in close proximity. Wherever the two cones would intersect each other, there a pollution is inevitable. But, even where they do not cross each other, there is danger, for the cone of soakage might strike a water-bearing stratum, and liquid impurities may thus be carried to great distances, polluting springs and causing zymotic disease.

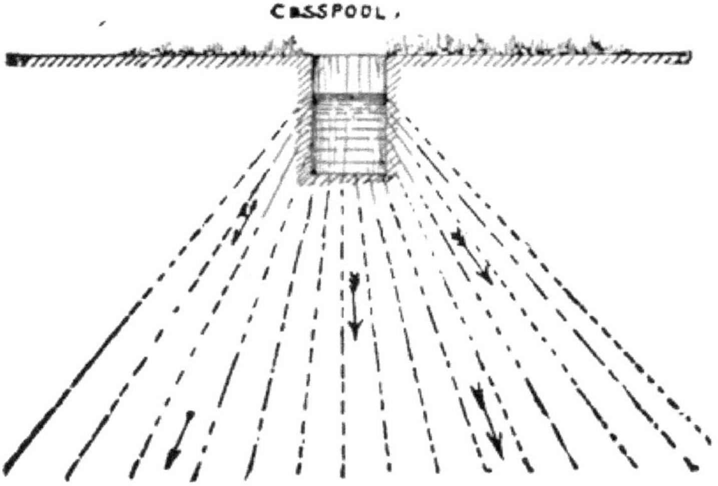

FIG. 181.—Cone of pollution.

A *leaching* cesspool is, under all circumstances, an abomination. Less dangerous and hardly as objectionable is the accumulation of household wastes in a *tight* receptacle or cesspool; but the latter should be built with particular care, made thoroughly secure against leakage, located as far away as possible from the dwelling, and efficiently ventilated. It should be of small dimensions, and consists better of two compartments, the first of which retains the solids and must be frequently emptied, cleaned and disinfected; while the second larger compartment holds the liquids, which must be disposed of on the ground, on the lawn, or in the vegetable garden, by frequent pumping out.

It is a mistaken notion, frequently met with in rural or suburban districts, where water-closets are not used, that slop-water from bedrooms and kitchens cannot, *per se,* become a dangerous nuisance. Hence, such liquid wastes are often disposed of by running them in open street gutters to the nearest pond or brook; they are also frequently dumped upon the ground around the dwelling, especially near kitchen windows. The emanations from a farmer's back yard on a hot summer's day are generally extremely nauseating and unwholesome.

But this is not all. Another not less dangerous nuisance is the common privy, which is still to be found in many cities, and is the rule in villages and isolated dwellings having no general water supply.

The prevalent form of privy is nothing but a large hole in the ground, a few feet deep, over which is erected the simplest kind of a shed, provided with a rough seat with hole. Who has not, on a hot summer's day, when compelled to pass near such privy, felt the offensive and truly sickening influence of the vile emanations from such an accumulated mass of excrement? Indeed, it is not surprising that we hear so much now-a-days of malaria and fever in the country.

Furthermore, I wish to enter earnest protest

against a method of house-drainage extensively practiced in many cities and towns, notably in Philadelphia, and in St. Louis. It is usual, in the case of smaller houses, to have inside the house, in the rear extension, only a kitchen sink on the first floor, and a bath or bowl on the second floor. In the rear of the yard a vault is built, over which a privy is erected. This vault is provided with an overflow or connection to the street sewer. Into it runs a waste pipe from the kitchen sink, which also receives the rain-water from the whole or a part of the roof. The excrement which accumulates in the privy vault is supposed to be washed out into the sewer with the flush from a good rainfall, but such is not the case—at least the flushing out is not a thorough one; flushing these vaults from a yard hydrant by means of a hose is most always neglected; frequent stoppages between the vault and the sewer, or further on in the street sewer, occur. The privy vault is seldom built thoroughly tight, consequently there is danger of soil pollution; and even where it is tight there is always a poisoning of the atmosphere of the rear yard with vile stenches. Such a privy vault is not much better than a common privy, and should not be tolerated by the authorities.

An English sanitary engineer truly said:

"An open privy cesspool is, in most cases, a nuisance. The addition of small quantities of water to effete organic matter causes fermentation and the liberation of the gases of decomposition, and, therefore, *all such matter should either be washed away with plenty of water, or water should be wholly excluded from it. Either an abundance of water or none at all is alone safe in this case.*"

Wherever cities have adopted the "water-carriage" system, the use of some kind of water-closet apparatus, which is vastly superior in point of comfort, decency and cleanliness, should be made imperative.

I cannot conclude this chapter better than by quoting the admirable words of Dr. Simon on forms of filth, producing disease:

"There are houses, there are groups of houses, there are whole villages, there are considerable sections of towns, there are even entire and not small towns, where general slovenliness in everything which relates to the removal of refuse matter, slovenliness which in very many cases amounts to utter bestiality of neglect, is the local habit: where, within or just outside each house, or in spaces common to many houses, lies for an indefinite time, undergoing fœtid decomposition, more or less of the putrefiable refuse which house-life, and some sorts of trade-life, produce: excrement of man and brute, and garbage of all sorts, and ponded slop-waters: sometimes lying bare on the common surface; sometimes unintentionally stored out of sight and recollection in drains or sewers which cannot carry them away; sometimes held in receptacles specially provided to favor accumulation, as privy-pits and other cesspools for excrement and slop-water, and so-called dust-bins, receiving kitchen refuse and other filth. And with this state of things, be it on large or on small scale, two chief sorts of danger to life arise: one, that volatile effluvia from the refuse pollute the surrounding air and everything which it contains; the other, that the liquid parts of the refuse pass by soakage or leakage into the surrounding soil, to mingle there, of course, in whatever water the soil yields, and in certain cases thus to occasion the deadliest pollution of wells and springs. To a really immense extent, to an extent indeed which persons unpracticed in sanitary inspection could scarcely find themselves able to imagine, dangers of these two sorts are prevailing throughout the length and breadth of this country, not only in their slighter degrees, but in degrees which are gross and scandalous, and very often, I repeat, truly bestial."

. . . . . "While it cannot be denied that ill-devised and ill-managed water-closets and their accompaniments have caused filth-diseases to a very large extent, a far larger range of mischief has attached to the other kinds of privy-arrangements: and of all the filth-influences which prevail against human life, privies of the accumulative sort operate undoubtedly to far the largest extent."

"The intention, and, where realized, the distinctive merit of a system of water-closets is, that in removing excremental matters from a house it does so with perfect promptitude, and in a perfectly neat and complete manner, not having any intervals of delay, nor leaving any residue of filth, nor diffusing any during its operation; and where the water-system is not in use, these objects ought still as far as possible to be secured. Thus, in the absence of water-closets, evidently any reasonable alternative system ought to include the following two factors, brought into thoroughly mutual adjustment: first, proper catchment apparatus in privies; and secondly, proper arrangements for privy scavenage."

. . . . . "Now, hitherto, in places not having water-closets, the general practice has flagrantly contravened those conditions. Either it has had no other catchment-apparatus than the bare earth beneath the privy-seat, and has trusted that this (receiving the excrements and often also the house-slops on to its natural surface or into a hole dug into it) would absorb and drain away the fluid filth, and serve during months and years as heaping-place for the remainder; or else it has had, as supplement to the privy, a large inclosed midden-stead or cesspool, partly or entirely of brickwork or masonry, intended to retain large accumulations of at least the solid filth, with or without the ashes and other dry refuse of the house, and in general dividing its fluid between an escape-channel, specially provided, and such soakage and leakage in other directions as the construction has undesignedly or designedly almost always permitted. Privies, such as these, have not been meant to have their filth removed except when its mere largeness of bulk (exceeding or threatening to exceed the limits of the privy-pit or cesspool or midden) might mechanically make removal necessary, or else when there might happen to arise an agricultural opportunity for the stuff; and public scavengering in relation to such privies has either had no existence, or has been adapted to the supposition of an indefinite local tolerance of accumulation. All this accumulation, with its attendant exhalation and soakage, and at intervals the shoveling and carting away of its masses of fœtid refuse, and the exposure of the filth-sodden catchment surfaces of privy-pits and middens, has been, as needs hardly be said, an extreme nuisance to those in whose vicinity it has been; and sometimes with the aggravating condition that, because of the situation of the privy, each filth-removal must be through the inhabited house. What nuisance this system at present constitutes in innumerable populous places, including some of our largest towns, can indeed hardly be conceived by persons who do not know it in operation; and the infective pollutions of air and water-supply, which it occasions to an immense extent in towns and villages throughout the country, are chief means of spreading in such places some of the most fatal of filth-diseases."

We have dwelt at some length upon the common defects in the drainage and sewerage of buildings. The closing chapters of our volume will be devoted to a brief description of what may be called the *chief features of a well-devised and well-constructed system of house drainage.*

We will arrange our subject under the following headings, viz.:

DRAINAGE OF DWELLINGS;

INTERNAL SEWERAGE OF DWELLINGS;

DESCRIPTION OF PLUMBING FIXTURES;

GENERAL ARRANGEMENT AND CARE OF FIXTURES;

EXTERNAL SEWERAGE;

DISPOSAL OF HOUSEHOLD WASTES.*

---

*The subject is treated at greater length and somewhat more in detail in the author's book, "House Drainage and Sanitary Plumbing," published by D. Van Nostrand, N. Y. (Second revised edition in press).

CHAPTER X.

SYSTEM OF INTERNAL SEWERAGE AS IT SHOULD BE IN A DWELLING.

Drainage of Dwellings.

THE term "drainage" in distinction to "sewerage" is intended to apply to the removal of sub-soil water from the site upon which the dwelling is to be erected. No house can be considered perfectly healthy unless provision has been made for carrying away any excess of moisture from the soil. Damp and wet cellars have a well-known influence in predisposing people living in such houses to pulmonary diseases.

To carry off sub-soil water, tile drains (common round land drains) should be laid at a depth well below the cellar floor, in parallel lines, their distance depending largely upon the character of the soil. The tiles may be 1½ or 2 inches in diameter, and should be laid with open joints, well wrapped with tarred paper or strips of cotton rags, to prevent dirt from falling in at the joints. Such branch drains should all be collected into one main cellar drain, of 2 inches diameter or of larger size, where the amount of water to be removed should be excessive, on account of springs, or for other reasons. If the house is a country house, this cellar drain can generally be continued to some low point, a ditch, ravine or a water-course, into which it should deliver. The outlet should be built in stone, or masonry, well protected by a strong grating, to prevent the entrance of rats or vermin. Where the water-course is subject to back-flooding from sudden rains, or some other cause, it may be necessary to apply to the outlet drain a tidal flap or ball valve. (See Figs. 40 and 87).

If the house stands on a city lot, the only outlet generally available is a sewer in the street. To connect such cellar drain directly to a sewer or to a house drain leading to it would be to lay sewer gas on to the house. There should be a *thorough disconnection* between a sub-soil drain and a sewer, which can be effected by a trap with a very deep seal, not liable to be affected by evaporation in

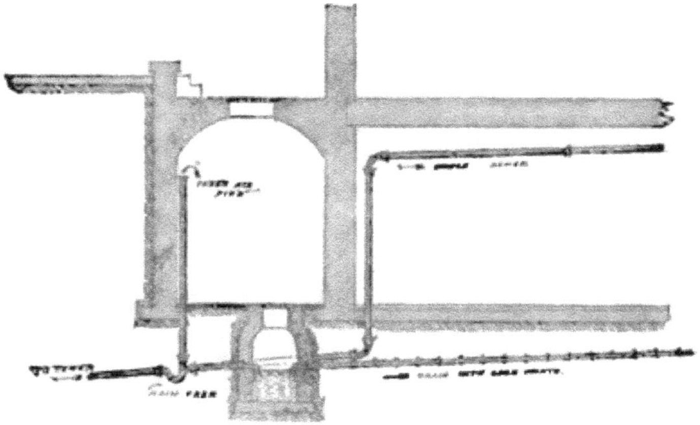

FIG. 182.—Sub-soil Drainage for City Dwellings.

hot, dry weather (Fig. 182), or by a gravel trap with an overflow to the sewer.

To ensure a dry and healthy cellar, it is necessary, in addition to the cellar drainage, to concrete the cellar floor, and to render it impervious to water by a good rendering of Portland cement, or better still, by a thin layer of asphaltum, spread on top of the concrete.

Dampness of foundation walls is equally bad and dangerous to health, and can be most efficiently prevented by asphalting the outer face of foundation walls to the surface of the ground, or by the use of double walls with air-space between, or better still, by an area built all around the foundation walls, well drained and well ventilated.

As it is not the purpose of this book to give general advice upon the construction of healthy houses—our subject being restricted to the *drainage* and *sewerage* of *dwellings*—we cannot enter more fully into a discussion of the important problem of how to prevent dampness in walls. The above hints, however, when followed, cannot fail to secure improvements in this direction.

Internal Sewerage of Dwellings.

All drain, soil, waste and air pipes inside of a dwelling (except the short branch wastes from fixtures) should be of iron.

The arrangement of soil and waste pipes must be as direct as possible, and long branch wastes under floors should be avoided everywhere.

Each stack should run up as straight as possible, avoiding offsets, which are objectionable.

None of the waste or vent pipes should be buried out of sight and rendered absolutely inaccessible. It is preferable to keep them in sight, except on the parlor floor. The public has long been accustomed and does not object to the running of steam pipes in plain sight, and there is no reason why soil pipes should not be treated in the same manner. Their outside can be painted, or, if desired, it can be gilt or bronzed, as is done with steam pipes. Where pipes must be placed in recesses or chaces in the walls, or in partitions, they should be covered with wooden panels, or boards, fastened with screws so as to be easily removed, should an inspection of the plumbing become necessary.

The soil, waste and air pipe system should be thoroughly tight, not only water-tight, but air-tight as well. Hence, the pipes must be of thoroughly sound material, and all joints must be perfectly made.

The system must be amply ventilated, and should have no "dead ends." Each soil pipe, therefore, must extend at least full size from the cellar to and through the roof; waste pipes must also be extended, but should be enlarged just below the roof to 4 inches in diameter, to prevent obstructions of the pipe in winter through hoar frost.

Wherever practicable, soil and waste pipes should run along a heated flue, as this will assist in creating an upward draft in the ventilating pipes.

The extensions above the roof should, in all cases, be not less than two feet high, so as to be well exposed to air currents; if near a chimney top, they must terminate well below it. In any case, soil pipe mouths should be located as remote as possible from ventilating shafts, chimney flues, or ventilating skylights.

The mouths of all pipes above the roof should be kept *wide open*. Return-bends are highly objectionable; ventilating caps clog up in winter time through hoar frost. None of the many patent ventilators are preferable to an open-mouthed pipe.* To prevent obstructions of the pipe, insert into the top a copper mushroom-shaped wire basket (commonly called leader guard).

Soil pipes should not be larger than four inches in diameter; vertical waste pipes for sinks or bowls, are generally two inches in diameter.

Each vertical line of air pipes must be at least two inches in diameter, increasing at the upper floors (in the case of high buildings) to three and even four inches diameter. Each line of air pipe should extend as straight as possible up through the roof, where its mouth should be well exposed to the wind, and provided with a grating, screen or basket for protection. Air pipes may, however, branch into their soil pipe above the highest fixture, thus avoiding a large number of holes on the roof. Each vertical line of air pipe must have the necessary fittings to connect the branch air pipes from traps to it. It is a mistake, frequently made, to use inferior material (lighter pipes) for such air pipes. They must, invariably, carry more or less foul gases, their joints should, therefore, be as tight as those of soil and waste pipes.

Leader pipes, if inside the walls, must be of cast

---

* Ventilators or exhaust cowls render, in some cases, excellent service, if placed on top of flues, to prevent smoky chimneys. No good reason exists for putting them on top of soil or air pipes.

iron or wrought iron, with thoroughly tight joints. If the leader opens at the top near attic windows, or near chimney flues or ventilating shafts, and if it is made of metal (galvanized iron or tin) and passes near windows of living or sleeping rooms, it must be trapped by a trap, with deep seal, located out of reach of the frost. Iron leaders with tight joints opening at the top remote from flues, or ventilators, or windows, should not be trapped.

Each stack of soil or waste pipe must have fittings in proper position to receive the flow from the fixtures. It is not absolutely necessary that the fittings on vertical soil pipes should be Y branches ; a Tee-branch, especially, if its flow line is shaped in a curve, will answer the purpose as well, and such is especially the case for small waste pipes joining the soil pipe.

The flow from all soil and waste pipes is collected in the cellar, the aim always being to concentrate the system as much as possible. As a rule, it is better to connect the rain water leaders with the drain that carries the waste water of the household. In country residences, where the rainfall is collected in a cistern, a separate system of pipes is required, and this is also the case, wherever the sewerage system of the city is the so-called "separate system."

The junction between upright soil, waste or leader pipes, and the horizontal drain, is of the greatest importance. The best support that could be given to it, is to build a brick pier under it, and to rest the weight of the upright pipe stack on it.

Sometimes a strong wooden post is of service, though not making as substantial a job as a brick pier. The junction should be made with an elbow fitting of a large radius, or with Y-branches and 45° bends, in order to make the change in the direction of the flow as gradual as possible. A right angled connection must not be tolerated, as it is sure to cause accumulation of soil and create stoppages.

It is to be recommended to run the main cellar drain *in sight* along one of the foundation walls, or to carry it along the cellar ceiling suspended from the joists or iron beams by strong iron hangers.

Where there are fixtures in the cellar, the main drain must run below the floor, and in this case it is advisable not to bury it entirely out of sight. Cleaning hand-holes should be provided at all junctions of branch drains with the main, also near or at bends, at the trap, and at the foot of vertical stacks. These hand-holes must be left accessible by building small man-holes with covers around them. Many authorities require every drain below the cellar floor to be laid in a trench with concrete bottom and with bricked walls, accessible throughout its entire length. This seems necessary only where inferior material is used, and where the workmanship is not first-class. With heavy iron pipes, tested not only at the foundry, but also after being placed under the floor of a house (by the water or air pressure test), and with few well-made joints it is better to bury the drain pipes in concrete, leaving out places for access only whereever really needed.

For all horizontal or inclined drains the rule should be laid down, that no junction should be made at right angles, with Tees; 45° Y or 67½° Y-branches must be used. All changes from the straight line must be made with curves of a large radius. It is advisable to use near junctions, curves and traps, hand-holes giving ready access to the drain pipes in case of accidental or malicious stoppage.

The fall required for the main drain will depend upon its size; the latter should not exceed six inches in most cases. Where a building is unusually large, it is better to have two main drains, each six inches diameter, than one drain of nine inches. As a rule, however, four and five-inch drains are ample for ordinary sizes of dwellings. Such a drain should, if possible, have a fall of ½ inch to the foot, but a fall of ¼ inch to the foot is sufficient to carry along whatever ought only to enter such pipes.

If the main drain is trapped, as is advisable in most cases,* the running trap of iron should be located just inside the cellar wall or else outside the house, in a man-hole. It should be located where it is not exposed to freezing.

In any case the trap must not be absolutely inaccessible, as it is possible that obstructions may occur at this point. The trap should, therefore, be provided with cleaning holes, closed air-tight with well fitting covers. It is advisable to run into this trap a leader, so as to insure its occasional flushing at each rain-fall.

*See Chap. IV., page 55.

To insure a full circulation of *fresh air* through the pipes, a fresh air pipe, of the full diameter of the iron drain, must run from just inside the trap to some point outside, well remote from windows, so as not to cause any objectionable smell, as it becomes at times—though seldom—an *outlet* instead of an inlet.

As soon as the main soil and drain pipe system is completed, its tightness should be tested by the water pressure test (see Fig. 22) or by the peppermint test.

Another equally reliable test is the air pressure test by a force pump and a manometer. In this test every part of the pipe system is subject to a *uniform* pressure, while in the water test the pressure increases with every foot of head of water, thereby often putting an unusually severe strain on the lower part of the pipe system.

The house drain should be of iron, to a point well beyond the foundation walls. Whether it should be continued any further with iron pipe, or whether vitrified, well-glazed pipes may be used for the external sewerage, will depend entirely upon the character of the soil. For made ground, *heavy* iron pipe is decidedly to be preferred, but care must be taken to lay the pipes on a good solid foundation. It is also safer to run a drain of iron pipes, where it passes near a well, furnishing drinking water. Occasionally roots of trees cause considerable trouble with vitrified pipe, especially if the joints of the latter are poorly cemented. In such a case, iron pipes, with caulked joints, are preferable.

Fig. 183 represents a section through a country dwelling, showing all drain, soil, waste and air pipes; also the fixtures and the mode of trapping these.

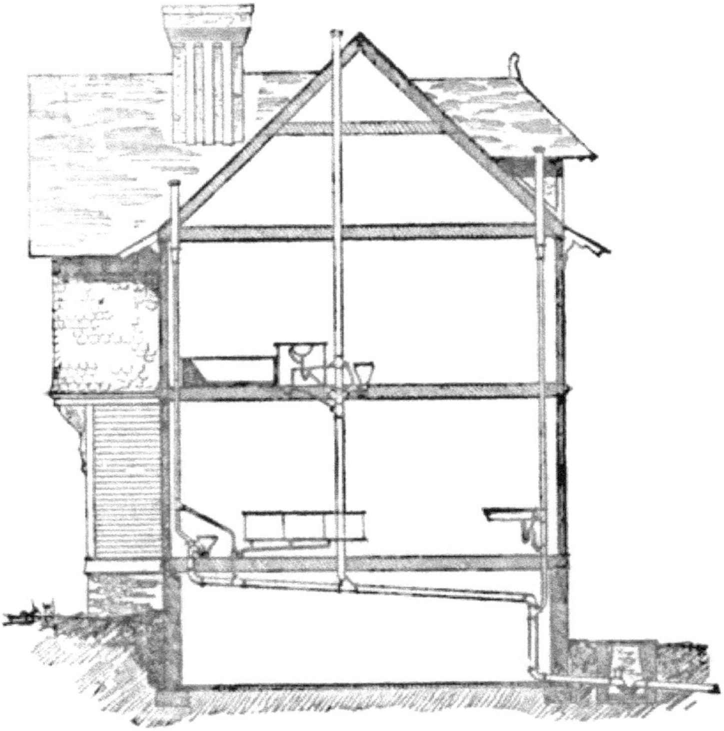

FIG. 183.—Soil, Waste and Air Pipe System in a Country Dwelling.

We will consider the external sewerage further on, and must now refer once more to the inside system of drain, soil, waste and air pipes, to discuss the materials best adapted to such a system.

For all drains laid underground or below the cellar floor, the best available material is *cast-iron.* In Chapter III. we have discussed the faults of common plumbers' pipe. The cast-iron pipe, fit for purposes of house drainage, must be equivalent in strength and weight to heavy gaspipe in order, first, to secure a rigid and strong line of pipe, and, second, to obtain air and water-tight joints.

To quote from Capt. Douglas Galton, an experienced engineer :

"The use of cast-iron for house drains, if the cast-iron is solid, sound and free from porosity, will prevent leakage and sub-soil tainting beneath the house, and will be as cheap as earthenware pipes in many cases." .... "Lead joints can only be made in a strong iron pipe, and the use of these joints is, to some extent, a guarantee of soundness, but every pipe should be carefully tested by water pressure, to see that there are no holes or flues."

Pipes of heavy cast-iron are generally manufactured in lengths of 12 feet, with a hub and a spigot end. As regards the strength of such pipes much will depend upon their manufacture. The metal used should be a re-melted pig-iron of homogeneous texture, free and easy flowing when poured into the mould ; the fracture must show a dark gray color. Then again, great care and diligence must be bestowed upon the making and drying of the pipe moulds and cores ; the loam and sand should be carefully chosen, in order to form smooth and substantial moulds.

It is now pretty well understood by all manufacturers of first-class cast-iron pipe that the pipes should be cast over end, in order to obtain a *uniform* thickness of shell, which is the great desideratum for all pipes. If cast in a lying or inclined position, the molten metal poured into the mould has a tendency to float the core and bend it upwards in the centre, consequently the thickness of the shell will be much greater at the lower part of the pipe.

Experts disagree in regard to the position of the socket while casting. In England, it is customary to cast heavy pipes with the socket downward. In such a position, it is claimed, the head of pressure of the fluid metal, equivalent to the length of pipe, will secure a strong socket, free from air bubbles or other defects. The top end will often be spongy, containing floating dirt, slag, scoriæ and air bubbles. Should this occur with the socket or bell end of the pipe, it would render the socket weak and often worthless for caulking purposes, while the spigot end may be cut off, if necessary. A large dead bead is often given to the spigot end of pipe, which is afterwards removed by cutting. In Germany, on the other hand, the custom prevails of casting pipes socket upward. In the United States of America, the general practice is to cast all pipes from 3″ to 12″ diameter, socket upward, while larger sizes are always cast socket downward. There are, I believe, practical advantages, such as easier drawing and removing of patterns, which influence the American foundries to cast with socket upward.

After being cast, all pipes should be carefully protected from sudden chills, the cooling should be gradual and slow, so as to avoid imperfections in the metal. After cooling off, the pipes are carefully cleaned with steel wire brushes and scrupulously inspected.

All such pipes should be straight, truly cylindrical, of a uniform thickness, of a uniform and homogeneous texture, of *perfect smooth surface,* free from flaws or other imperfections, and the spigot end should truly fit the hub of the pipe. The pipe must not be brittle, but must allow of ready cutting, chipping, drilling or threading.

The thickness of metal of cast-iron pipes, used for sewerage purposes, should be about as follows:

2 inch pipes	$\frac{5}{16}$	inches	thick.
3 " "	$\frac{5}{16}$	"	"
4 " "	$\frac{3}{8}$	"	"
5 " "	$\frac{7}{16}$	"	"
6 " "	$\frac{7}{16}$ or $\frac{1}{2}$	"	"

After a careful and thorough inspection, each pipe must be tested under pressure in a hydraulic testing machine. Whilst under such pressure, the whole length of the pipe should be repeatedly struck hard with a heavy hammer, in order to detect flaws or weak parts of the pipe shell. If the sound of the hammer striking the pipe metal is clear or bell-like, it is a pretty sure indication of the absence of any of the above imperfections.

If it is thus made sure that the pipe is free from air-bubbles, flaws, shrinkage-cracks, sand-holes, etc., the pipe must be coated in order to protect it against corrosion. The best solution known for cast-iron pipe is Dr. Angus Smith's patent coal-tar varnish. After placing the pipes in an oven, they are heated so as to well open the pores, and the solution is likewise kept hot in a tank, care being taken that it does not get too great a consistency. The pipes are then immersed in a bath for about 15 to 20 minutes, then removed, when the surplus of varnish is allowed to drip off from the pipe. There should be only a thin smooth coating of varnish over the pipe. The pipes are now ready for use.

A word of caution seems appropriate in regard to failures of cast-iron pipes through rough handling during carting, or loading and unloading. Great care must be taken not to throw any pipes violently on the ground, nor to expose unloaded pipes to violent accidental blows. Owing to the brittleness of the material, cast-iron pipes often split or break off at the ends, and the split, although hardly perceptible on the outside, may continue longitudinally very far, which fact can only be detected by the pressure test.

In laying such cast-iron pipes, the spigot end of one pipe must be inserted as straight as possible and concentric into the hub end of the next pipe, care being taken before doing so that the pipe is clean and free of all dirt on its inside. A gasket of oakum or dry hemp is then inserted into the space between socket and spigot, and well rammed with a caulking tool. This gasket should fill about one-half of the depth of the hub, its object being to prevent any molten lead from flowing into the pipe at the joint, also to assist in tightening the joint.

A roll of good, tough clay is placed around and pressed against the front of the pipe bell with an opening on the top, where the two ends of the clay roll meet, large enough to admit of pouring in the lead. This clay ring prevents the escape of molten lead while running the joint.

The lead used for making pipe-joints should be

soft and pure, without any admixture of tin, zinc or other metal If hard or impure lead is used, the caulking operation strains the bells often so much as to burst or split them. The lead is melted in a large pot kept on a furnace. It should be kept at a proper temperature in order to prevent too sudden cooling while pouring it. A large ladle (Fig. 184),

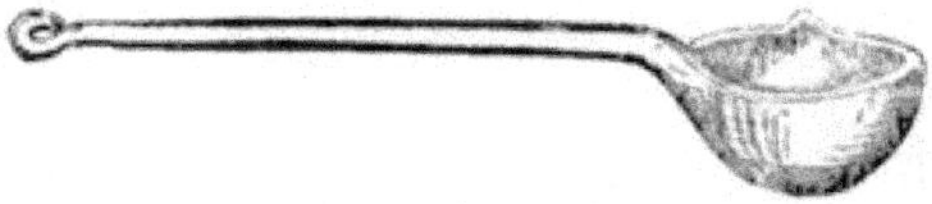

FIG. 184.—Ladle.

which must be capable of holding enough molten lead for one joint, is used to pour the lead into the space between spigot and bell. It is important that enough molten lead is poured in at one operation to quite fill the joint, for if the lead is not poured in a continuous stream the joint will not be perfect and homogeneous.

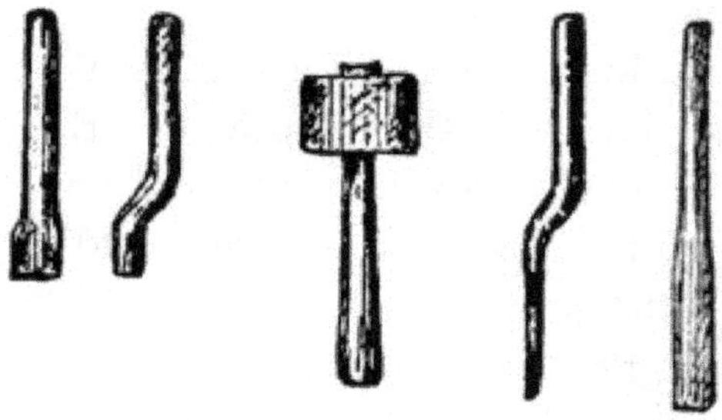

FIG. 185.—Caulking Tools.

As soon as the socket is quite filled, the ring of clay is then removed and the lead allowed to cool, while the superfluous lead is cut off with a cold chisel. The lead naturally shrinks and would not, *per se*, make a tight joint, but requires a thorough setting up or caulking, which is done first with a hammer and flat caulking tool, next with a similar broader tool, with a slight curve corresponding to the size or radius of the pipe. Fig. 185 shows the caulking tools generally used.

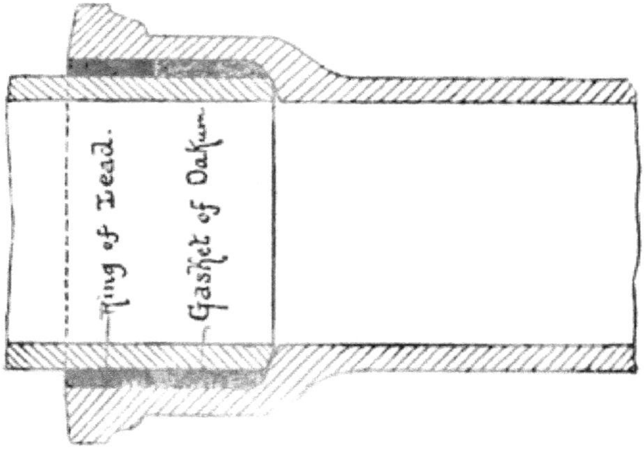

FIG. 186.—Finished Caulked Joint.

Fig. 186 shows the caulked joint in section, when finished. To insure a perfect joint the ring of lead should have an equal thickness all around. This thickness varies from $\frac{1}{4}$ to $\frac{3}{8}$ inch ; the ring should have a depth of from $1\frac{1}{2}$ to 2 inches. The following table, showing the amount of lead required for a joint, may serve as a guide:

DIAMETER OF PIPE.	WEIGHT OF LEAD IN LBS.	DEPTHS OF LEAD IN INCHES.
2	2	$1\frac{1}{2}$
3	$2\frac{3}{4}$	$1\frac{5}{8}$
4	4	$1\frac{3}{4}$
5	$5\frac{1}{2}$	$1\frac{7}{8}$
6	7	2

The lead must be left exposed, so as to show the marks of the caulking tools. No paint, cement or putty should be used to fill the space in front of the caulking ring.

A proper caulking operation always puts a heavy strain on the sockets of the cast-iron pipe, and, in order to withstand it and prevent the bursting of bells, the latter should be designed very strong, with an extra thickness of metal at the end of the hub and at the point where the socket joins the pipe. The thickness of the hub should not be in direct proportion to the thickness of the pipe. This latter point is less thoroughly understood. The failure of common light plumbers' pipe is largely due to the fact that both the pipe shell and the thickness of the bell are reduced to a minimum. While it may be possible to reduce the former slightly, wherever there is no heavy inside pressure or outside superincumbent weight, the latter should always be kept heavy enough to withstand a thorough caulking.

In making lead joints in cast-iron pipes, much, of course, depends upon the skill, sound judgment, experience, but above all, upon the honesty of the workman. Careless or dishonest mechanics are very apt to do the caulking of the lead imperfectly, to omit this operation entirely at the under side of the joint, which is difficult of access, and not so readily inspected. Perfect workmanship is absolutely essential in the case of iron drain pipes, not less than for water or gas mains.

Cast-iron pipes are, sometimes, cast with flanges at both ends, but such joints are more difficult to make tight, and a flange joint, being more unyielding than a lead joint, causes fractures of the pipe. Such pipes are not used for sewerage purposes.

There are other joints in cast-iron pipe, such as the turned and bored joint, and many patent joints, the consideration of which would lead us too far.

We must, however, mention *rust joints* in cast-iron pipe, as these are frequently used, especially for blow-off pipes from boilers, and for steam pipes. In place of lead, an iron cement is employed to make a joint. A quickly-setting cement is composed of :—

98 parts fine cast-iron borings.
 1 part flowers of sulphur.
 1 part sal ammoniac,

to be mixed with boiling water before use.

A slow-setting cement is made as follows :—

197 parts fine cast-iron boring.
 1 part flowers of sulphur.
 2 parts sal ammoniac.

Cast-iron drain pipes require a number of fittings, such as elbows, Y branches, traps, Tee branches, which are also cast with bells and connected to the pipes in the same manner as lengths of pipe are put together.

All such castings should be carefully examined before use. They should be sound, smooth, especially on the inside, without lumps, sand-holes, flaws or scoriæ. The inspection of castings is very important, for, as Mr. Baldwin Latham says :

"There are faults, to which all articles made of cast-iron are liable, and which may escape observation even after the most careful scrutiny, and, in consequence, there will ever remain a certain degree of uncertainty as to the strength of iron castings, for there are numerous instances which may, more or less, affect the quality of the manufactured article, such as unequal contraction in cooling, imperfections from latent flaws which may be concealed by a covering of sound metal, the brittle nature of the material, the presence of some deleterious agent in the metal itself, all tending to render cast-iron more or less uncertain, and liable to fail without warning.....The proper admixture of the iron in the foundry is one of considerable importance in order to ensure a perfect casting ; for, as different varieties of iron have different points of fusion and varying rates of cooling, unless a proper admixture is ensured, the casting will have within itself an element tending to produce its own destruction, for, while some of the metal may be in perfect fusion, other parts may be imperfectly fused, while again others may be burned; or in cooling, some of the metals may cool faster or slower than others, consequently the casting may be thus brought into a state of unequal tension, or, as it is technically termed, 'hide-bound,' when such slight influences as sudden change of temperature may lead to its instant destruction."

If cast-iron pipe is used for vertical soil, waste and vent pipes, it should be of the same character and quality as above described for drain pipes, for only pipes of such superior properties allow the construction of a pipe system, equally tight as regards leakage of sewage and leakage of sewer gas.

It seems doubtful if cast-iron pipes of such greater lengths are equally adapted for use in vertical as in horizontal positions. We may also question the practicability of using pipes with such heavy bells for upright pipes, as the bells would not only present an objectionable appearance where pipes are kept in sight, but would occupy considerable space in partitions and chaces.

On the other hand, we deny the objection often made against iron pipes for soil pipes, that the inside is apt to corrode, for as Mr. Julius Adams, a civil engineer of experience, has said:

"It has been found that the objection urged against iron soil pipes—that they are liable to rust out from the inside—is invalid, since they soon become coated with a greasy film, which entirely prevents corrosion."

Within the last few years *wrought-iron* has been used extensively for soil, waste, air and leader pipes. Although architects have, in a few instances, used wrought-iron pipes for such purpose, in connection with common steam-fittings, we refer in particular to the work known as the Durham System of House Drainage, and as this presents novel features of interest to engineers, architects and sanitarians, we will give a full description of it.* Its chief departure from the common system of plumbing consists in the use of *wrought-iron pipes* for all pipes *above ground*, especially for all upright, soil, waste and vent pipes.

The pipe used is the standard, wrought-iron,

* At the time of writing this volume, the author is connected with the Durham System of House Drainage, holding the position of Chief Engineer of the Durham House Drainage Company, of New York, which constructs the above system. The author has endeavored to give a fair and impartial description of this system of construction, leaving the decision as to its merits and superiority over the common system to the disinterested reader.

lap-welded steam pipe. This is extensively manufactured by many "tube works" at the rolling mills in lengths of about 20 feet. Bars of wrought-iron of a width corresponding to the circumference of the pipe are bent up by means of powerful machines, while in a red heat, to a circular shape. The ends of the smaller sizes (up to 2 inches diameter) are made to butt against each other, while the larger sizes lap over. The bars are then again highly heated and welded together, after which operation they are adjusted so as to be exactly circular in shape.

Before leaving the works, and while hot, the wrought-iron pipes are immersed into a tank, containing hot liquid asphalt, which coating of the pipes effectually protects their inside against corrosion. All standard wrought-iron pipes are tested at the works by hydraulic pressure up to 500 lbs. per sq. inch, and a guarantee of good and durable material is thus secured.

The following table exhibits the size, thickness and weight of pipes, used for soil and vent pipes:

Size of Pipe.	Thickness of Pipe.	Weight in lbs. per ft.
2"	.154	3.67
3"	.217	7.55
4"	.237	10.73
5"	.259	14.56
6"	.280	18.77

Such pipes are put together same as the steam pipes, with *screw joints.*

The screw thread, cut externally on the pipe, is slightly tapering, and so is the internal thread cut on the fittings. It is customary for pipes from 2 to 6 inches diameter to have 8 screw threads per inch. These threads were formerly cut on a lathe, if done by machine work; if by hand, by the use of die-stocks. Since a number of years large hand and power pipe-cutting machines have been manufactured, which use dies and cutters, by which a large saving in time may be effected. Instead of cutting internal threads of fittings in a lathe, they are now tapped by powerful tapping-machines.

Wrought-iron pipes are screwed into couplings, or fittings, by means of pipe-chain tongs (see Fig. 187), on which a man can exert a powerful leverage, thus securing the great desideratum, *tight* joints.

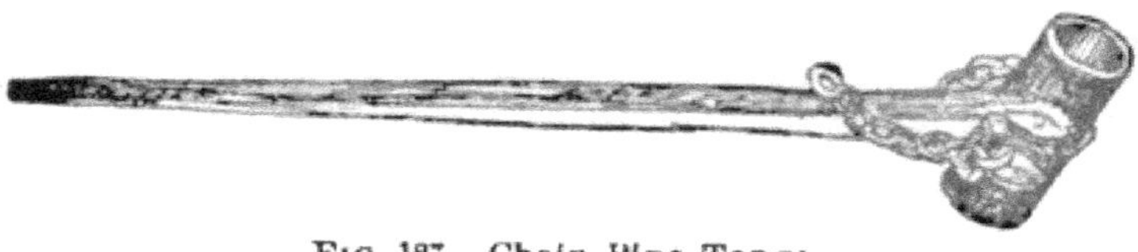

FIG. 187.—Chain-Pipe Tongs.

In order to make up for imperfections in the threads of the pipe and fitting, a paste is used in making the joints, consisting of an equal mixture of white lead and linseed oil with red lead. This paste hardens after some time, and forms a tight packing in the screw-joint.

The pipes are cut to required lengths from exact measurements, in a power pipe-cutting and threading machine. Straight lengths of pipe are screwed together by means of wrought-iron couplings; for changes of direction, special fittings, such as elbows, T and Y branches are used, which will be described later. It is generally possible to run such soil and vent pipes from floor to floor without intermediate joints; the total number of joints in each soil pipe stack is consequently largely reduced.

We quote from C. W. Durham, C.E., the inventor of this soil pipe construction:

"*Proper mechanical* construction must be the foundation of any good system of drainage."…. "By the use of wrought-iron pipes and screw joints we construct a 'drainage apparatus' within the building, which is gas and water-tight as regards the joints; rigid, yet elastic; entirely independent of walls or floors for support, and absolutely invulnerable. As a structure, it will last as long as any building will stand—*without any outlay for repairs*."

…."When lengths are screwed together in a wrought-iron coupling, the joint is as strong as any other part of the pipe, and they will stand up vertically, from a solid base, to the height of any building without lateral support."

…."The result attained is a system of pipes which are independent of the building for support; which cannot be cracked or broken; and whose joints are permanently gas-tight beyond the shadow of a doubt."

Fig. 188 shows a soil pipe stack as constructed and erected by the Durham Company.

A is the running trap on the main house drain, with cleaning holes and a fresh-air pipe, B, on the house side of the water seal, carried to a point remote from windows, or to the street curb. C is the soil pipe elbow on which the whole straight soil pipe stack, FF, is erected. E is a plug in the soil pipe ell to remove any obstruction at this bend. D is a 4x2 Y branch for a sink waste, G a 4x2 double Y branch for bath and bowl wastes; H is a double cross with grade, and branches LL from water closets.

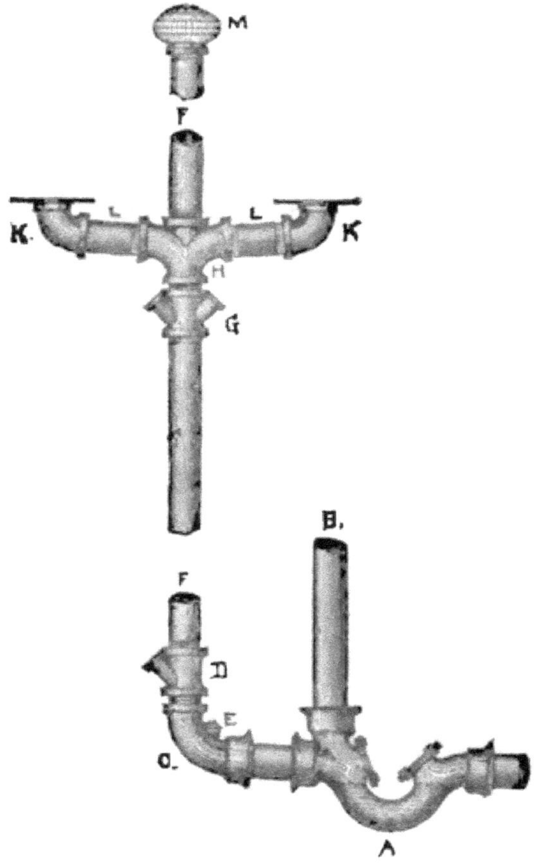

FIG. 188.—Soil Pipe Stack as erected by the Durham House Drainage Company.

A novel feature of this construction is also the manner of supporting the water-closet, the chief plumbing appliance in a dwelling, directly from the

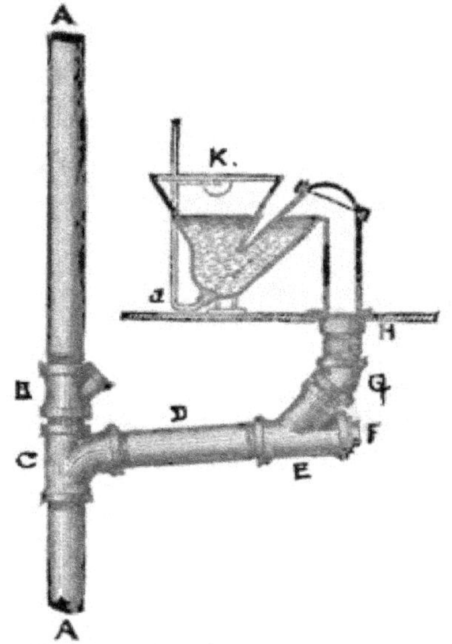

FIG. 189.—Soil Pipe with branch to water-closet, having trap above the floor.

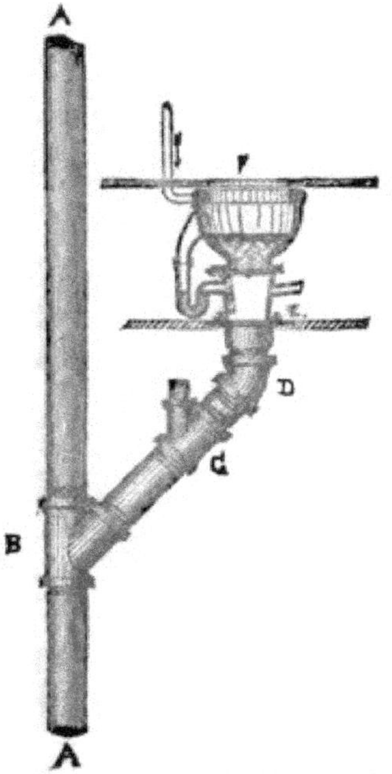

FIG. 190.—Soil Pipe with branch supporting a trapless water-closet.

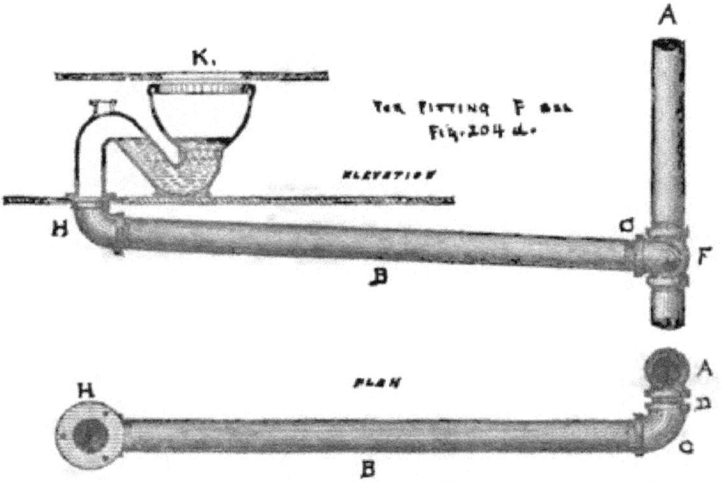

FIG. 191.—Soil Pipe with branch supporting a closet at a distance from the soil pipe.

soil pipe, independent of the floor, on special elbow fittings or water closet flanges K K. This is especially important for water-closets with trap above the floor, as in the common system the joint at the floor is liable to open through settlement of floor or joists. Fig. 189 shows a water-closet with trap above floor, supported on the branch from the soil pipe. Fig. 190 shows a trapless closet supported in

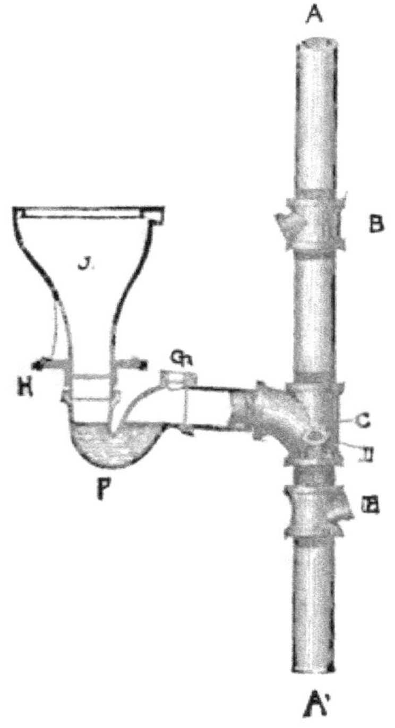

FIG. 192.—Soil Pipe with branch supporting a hopper-closet, with iron trap below the floor.

the same manner. Fig. 191 shows how closets at a distance from the soil pipe are supported on a cast-iron base, to which they are securely attached. This cast-iron base rests on the floor, but its waste pipe is connected to the soil pipe by a *flexible* joint. A settlement of the floor and a subsequent sinking of the cast-iron base, holding the water-closet, cannot loosen or open the joint at the soil pipe. Fig. 192 shows the manner of supporting water-closets, requiring a trap *below* the floor.

Fig. 193 shows the manner in which the wrought-iron soil pipes are screwed into the elbow fittings

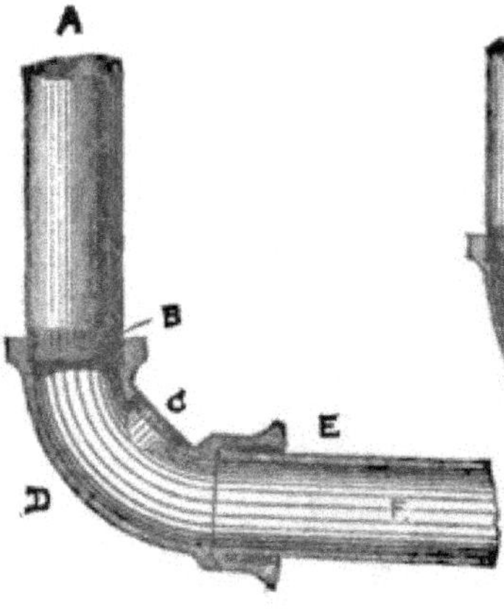

FIG. 193.—Soil Pipe screwed into soil pipe ell, and drain pipe caulked into the same.

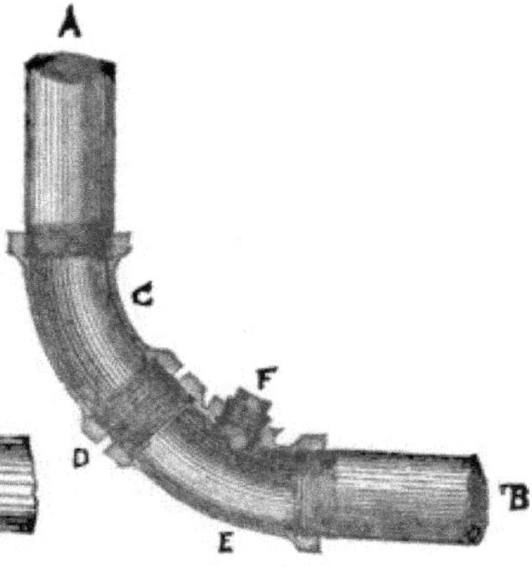

FIG. 194. — Junction between vertical and horizontal pipes, made in wrought-iron with easy curves.

which connect them with the underground drain. It will be noticed that the change in the direction of the flow is effected with an easy curve. Fig. 194 illustrates the same junction betwen soil pipe and drain where the latter is of wrought-iron and carried along the cellar wall. In both cases the construction provides for hand-holes closed by plugs for removal of accidental obstructions.

For drains under ground, the Durham System uses heavy *cast-iron gas pipe*, with lead joints, such as described before in speaking of the proper material for house drains. To quote from Mr. Durham:

"Lead joints and cast-iron drains are employed only for pipes in a *horizontal* position, in which there can be no *pulling* strains. All other joints are *screw* .joints, made with wrought-iron pipe."

The fittings of the Durham System are *special* fittings throughout. Common steam-fittings are unfit for purposes of house drainage, as they leave interior depressions, when the pipe is screwed up, which would collect sewage. (See Fig. 195a). The fittings for wrought-iron pipe in the Durham System are tapped with a shoulder (Fig. 195b),

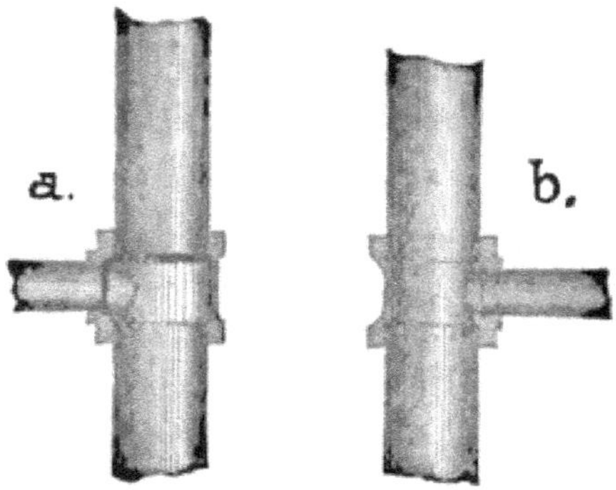

FIG. 195a.—Steam pipe screwed into a common steam fitting.

FIG. 195b.—Asphalted wrought-iron pipe screwed into a special fitting of the Durham System.

and when the wrought-iron pipe is screwed home, its interior and that of the fitting form a practically continuous line. (See Fig. 196). All pipes, however, do not screw up equally, and it happens that occasionally a small recess remains between the end of the pipe and the shoulder of the fitting. Such recess will be too small to be of any harm, at any rate it cannot collect more sewage matter than the inevitable recesses in cast-iron pipe at the point where the spigot end should touch the inside of socket. With *tight* joints, it would seem as if serious harm need not be apprehended from such unavoidable imperfections.

FIG. 196.—Screw Joint.

The usual fittings for heavy gas pipe, such as are used by gas and water-works, are not well adapted for sewerage purposes. The Durham System uses, therefore, a large number of special fittings for cast-iron pipe, and also some for both cast-iron and wrought-iron pipes. All these have hubs of great strength, shaped as shown in Fig. 197.

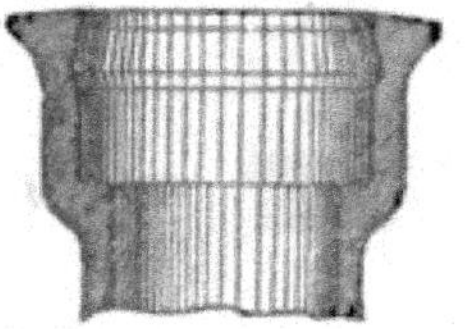

FIG. 197.—Socket of cast-iron special fitting of the Durham System.

All fittings are carefully protected from rust by dipping them in a bath of liquid asphalt.

The following is a condensed list showing the varieties of fittings, manufactured for use with the Durham System, a few of these being shown in sketches. It is evident that wrought-iron pipe requires a larger variety of elbows and other fittings, as the screw joint does not allow of the least deviation, such as may be made in the lead caulked joint. The special fittings are:

a, for cast-iron pipe, 3″, 4″ and 6″ diameter:

Running traps (Figs. 45 and 46), leader traps with deep water seal and cellar-floor traps (Fig. 167).

90°, 60°, 45°, $22\frac{1}{2}$°, $11\frac{1}{4}$°, $5\frac{5}{8}$° ells, with one or two hubs. (Fig. 198, *a*, *b*, *c*, *d*, *e*, *f*).

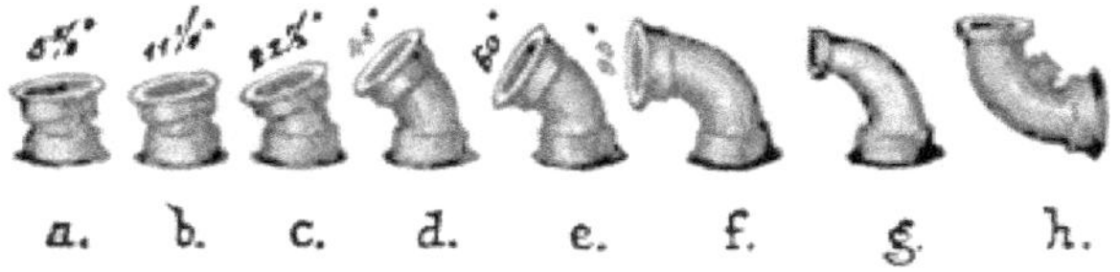

FIG. 198.—Elbow fittings of the Durham System, for cast-iron pipe.

Tee branches of all sizes. (Fig. 199, *a*, *b*).

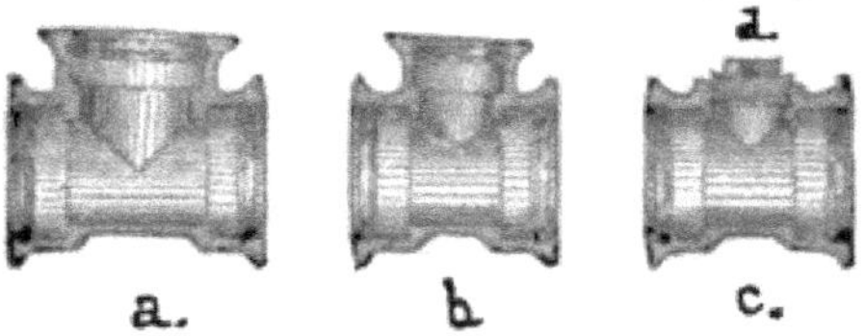

FIG. 199.—Tee fittings of the Durham System, for cast-iron pipe.

Y branches, 45° (Fig. 200), all sizes, with or without hand-holes.

Y branches, 90° (Fig. 201), all sizes, with or without hand-holes.

Reducers, for all sizes.

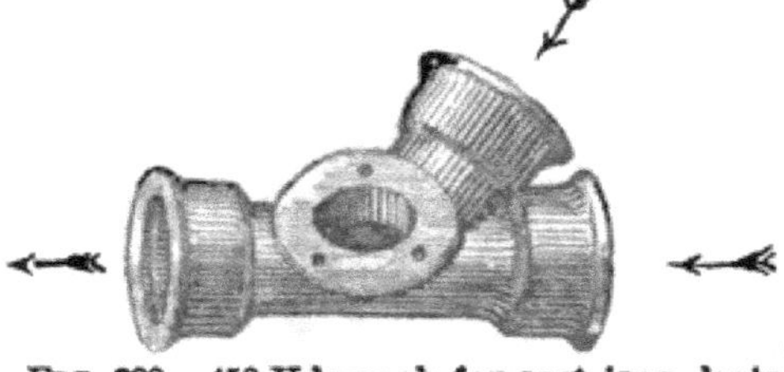

FIG. 200.—45° Y branch for cast-iron drain pipes, shown in perspective.

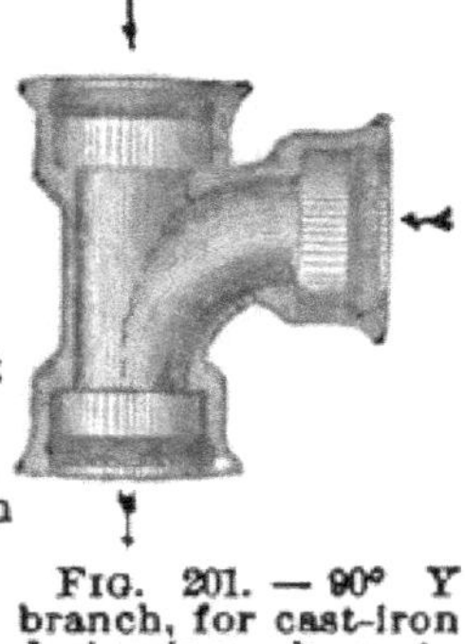

FIG. 201. — 90° Y branch, for cast-iron drain pipes, shown in horizontal section.

b, for cast-iron pipe, with openings tapped for wrought-iron pipe:

Soil pipe ells (Fig. 193 and Fig. 198, *h*).

Sink ells, all sizes. (Fig. 198, *g*)

Tees of all sizes. (Fig. 199, *c*).

Cast and wrought-iron connection pieces or couplings. (See Fig. 202, *a*, *b*, *c*).

FIG. 202.—Cast-iron and wrought-iron connection fittings of the Durham System.

c, for wrought-iron pipe, 2", 3", 4", 5" and 6" diameter:

Plain ells, ells with ½" and with 1" grade, Three-way ells, 60°, 45°, 22½°, 11¼° and 5⅝° ells. (Fig. 203, *a*, *b*, *c*, *d*, *e*, *f*).

FIG. 203.—Ells for wrought-iron pipe.

Water-closet ells (Fig. 203, *h*), and water-closet flanges

Plain Tees and graded Tees of all sizes (Fig. 204, *b*, *c*), water-closet Tees (Fig. 204, *a*), Crosses, with and without grade (Fig. 205), Increasers and Reducers of all sizes, 45° and 67½° Y branches, in all sizes (Fig. 204, *d*, *e*).

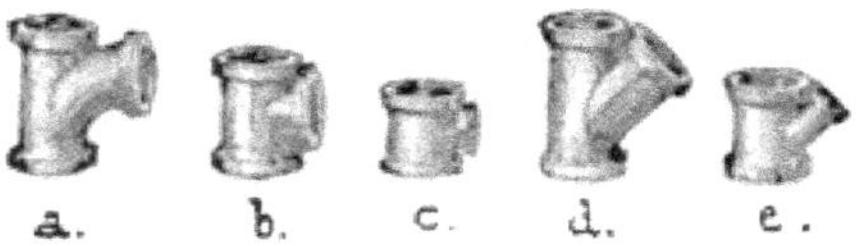

FIG. 204.—Tees and Y branches for wrought-iron pipe.

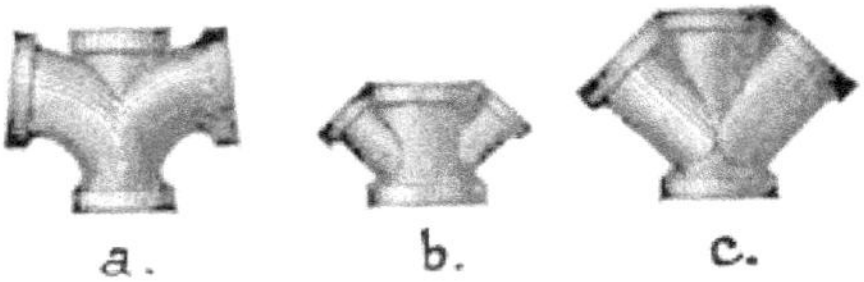

FIG. 205.—Crosses and double Y branches.

Water-closet traps (Fig. 192), yard drain traps, bushings, plugs, couplings, nipples, caps, union couplings and flange unions.

It occasionally happens in any system of soil or waste pipes that a length of pipe must be taken out and replaced, which can only be effected with plumbers' soil pipe by bursting a fitting. Such a result can also be attained in wrought-iron soil pipes by breaking a fitting ; the new length can be inserted either by a flange-joint, or else by the use of a running thread and a lock-nut. It must be remembered that, in the case of plumbers' soil pipe, a heavy knocking to break the pipe is likely to loosen many, if not all lead joints of the stack, while the screw-joints are not as easily affected.

It is true, on the other hand, that cast-iron pipes are more easily and quickly cut for making connections without the necessity of great mechanical skill or any expensive tools. Wrought-iron pipes require heavy and costly stationary machines to which the pipe must be sent, or else slow-working and expensive hand-tools for cutting and threading.

The lengths must be measured very accurately, and put together by skilled mechanics.

Some of the fittings, for instance Y branches for wrought-iron pipe, are not so easily put in place, on upright pipes in chaces, as the cast-iron fittings, but a skilled mechanic is generally able, with a little ingenuity, to overcome such difficulties.

It has been repeatedly asserted that wrought-iron rusts quicker than cast-iron, if plain and entirely unprotected. This is true and well-known to every engineer, but it does not prevent engineers from using an otherwise excellent, and, in many respects and for many uses, superior material. All iron pipes used for sewerage purposes must be efficiently protected against corrosion, and such is done with cast-iron pipes by coating them with coal-tar pitch, while wrought-iron pipes are dipped, thoroughly heated, into hot asphalt. There seems to be no reason why such coating, if done with equal care, should wear off, when the pipe is in use, quicker from wrought-iron than from cast-iron pipes. It is a fact that all soil and waste pipes are coated after a little use with a peculiar, greasy slime, which tends to protect the pipe—a cast-iron as well as a wrought-iron pipe—against corrosion.

As mechanical science advances, better means will undoubtedly be available to protect soil pipes from corrosion.

Amongst rust-preventing processes of recent origin I mention the *Bower Barff Rustless Process*, which consists in subjecting iron or steel to the action of superheated steam in a furnace, until the surface of the iron is covered with a more or less thick coating of magnetic oxide, which, as is well-known, is unaffected by exposure to air or moisture. The advantages of such a process are obvious: above all, every part of the article is reached and treated, while with painting, oiling, enameling, or asphalting, corner nooks and flaws in the iron may not be reached, thus not insuring such a thorough protection against rust.

The thoroughly jointed and ventilated soil and

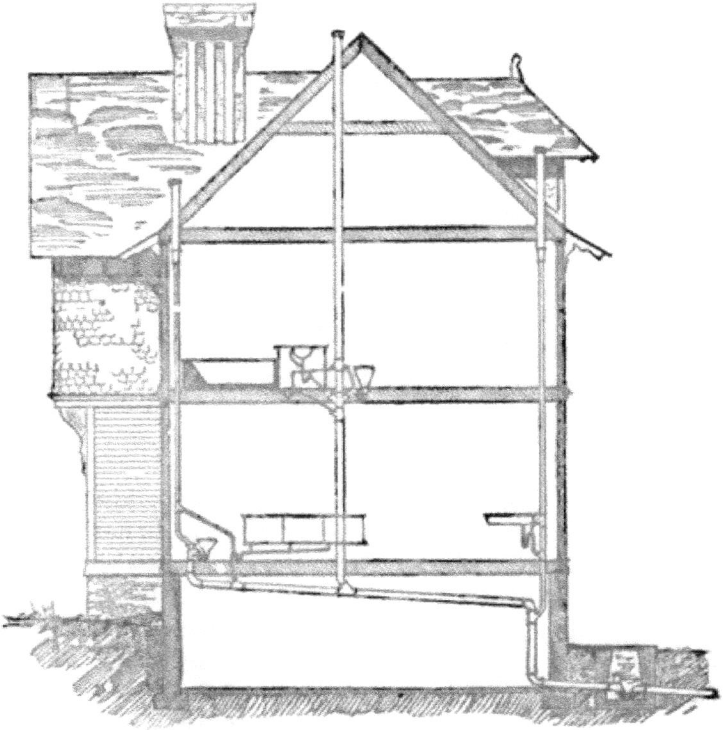

FIG. 183.—Soil, Waste and Air Pipe System in a Country Dwelling.

waste pipe system, receives, as shown in Fig. 183, the waste water from plumbing fixtures through short branch waste pipes of drawn lead pipe. The latter is made of all sizes in coils of any desired length, by pressing molten lead by means of a hydraulic press through dies, through which a core is inserted. Waste pipes of lead should be of the following sizes :

For one wash-bowl	1¼	inches	diameter.
For a row of basins	1½	"	"
For a bath-tub	1½	"	"
For a row of bath-tubs, likely to be used at once	2	"	"
For a pantry sink	1¼	"	"
For a kitchen sink	1½	"	"
For a set of laundry trays	1½	"	"
For a slop sink	1½–2	"	"

The weight should be about 2 lbs. for 1¼ inch pipe, 2½ lbs. for 1½ inch pipe, and 3 lbs. for 2 inch pipe. All joints in lead pipe should be wiped solder joints, and no cup joints should be tolerated, except where local circumstances render the wiping of a joint impossible. The following brief description of the manner of making a wiped joint is taken from the "Metalworker :"

"The first thing to do in setting about making a joint is to straighten the two pieces of pipe, to get out all buckles or sags. It is essential that the two pieces of pipe to be joined shall be so placed or fastened as to be quite firm during the operation, so that they shall retain their position while the joint is being wiped. It is also necessary that the ends of each of the pieces of the pipe shall be clean and free from oxide for a short distance back from the edges which are to be joined. In order to prevent the solder from adhering to other parts than those desired, both pieces of pipe are to be smeared with what is called plumbers' soil (composed of glue and lamp-black dissolved in water), as far from the joint as there is any liability of the solder touching. The soil is applied to the pipes with a small brush. The operation of making the ends of the pipe which are to be joined clean and free from oxide is to scrape them with a shave hook as far back as it is desired the solder should take effect. After the surfaces of the two pipes have thus been made bright, they are to be lightly rubbed over with tallow, in order to prevent oxidation by the air. Prior to the operation of scraping the ends of the pipe just mentioned, one of the two pieces is to be opened somewhat in order to let the end of the other

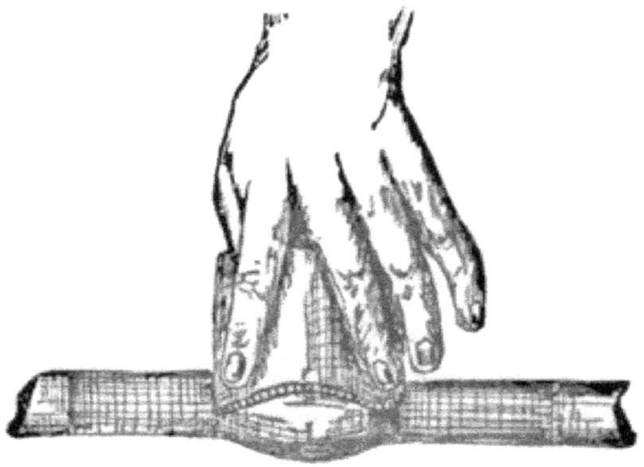

FIG. 206.—Wiped Joint in Lead Pipe.

piece fit into it. The opening of the pipe is accomplished by means of a turn-pin. After the pipes have been properly prepared as described, the two ends are placed together, and the joint is closed up by tapping it lightly with the hammer. At this stage the operator is ready for the solder, and to use the plumbers' cloth or soldering cloth. Solder is now poured on to the two ends of the pipe by means of the ladle. Pouring the solder against the joint, the plumber works it backward and forward, and around the joint, his fingers being protected all the time by the cloth until the pipe has become heated, and sufficient solder has adhered to it to make the joint. Having shaped it into a bulbous form, he completes the work by wiping it to shape after the manner indicated in Fig. 206."

Where lead pipe joins iron pipe the following mode of connection is recommended: If the soil or air pipes are of cast iron, brass caulking ferrules must be used, soldered to the lead pipe, and caulked with oakum and lead into the hub of the iron pipe. Where the soil pipe system is of wrought iron, lead waste pipes and lead branch air pipes from traps are connected to it by brass screw nipples, wiped to the lead pipe with solder, and screwed tightly with red lead and a wrench into the threaded opening of the fitting.

Each fixture, connected to the soil or waste pipe system, must be provided as near as possible to its outlet with a suitable trap, secure against siphonage, back pressure, evaporation, etc. If lead traps are used, the weight of the lead should be equivalent to the weight of the lead pipe.

Each fixture should, wherever possible, discharge into the main soil or waste pipe independently. The branch wastes should in all cases be as short and direct as possible, and this will largely depend on a judicious planning and locating of fixtures by the architect.

Overflow pipes, if such are used for fixtures, must connect to the waste pipe on the inlet side of the trap or below its water level, or else they must discharge over a "safe."

Drip pipes for safes under fixtures should not have any connection whatever with any soil or waste pipe or drain. They should be collected in the basement or cellar and discharge over an open sink, so that any leakage may be at once apparent.

Refrigerator wastes must never be directly connected to any soil, waste or drain pipe. These wastes are very apt to become coated with slime and dirt in a short time; they frequently stop up, and are generally liable to become offensive, especially if the ice used is very impure.

The outlets of all "set" fixtures except water closets should be protected against obstruction or chokage by a fixed strainer.

I close this chapter with the following brief description by Mr. E. C. Gardner, architect, of what a well devised soil pipe system should be:

"Theoretically, this is the whole machinery of safe, 'sanitary' plumbing: A large open pipe kept as clean and free as possible, into which the smaller drains empty, these smaller drains or waste-pipes having their mouths always full, and being able, so to speak, to swallow in but one direction. Everything can go down; nothing can come up. That all these pipes shall be of sound material, not liable to corrosion; that the different pieces of which they are composed shall be tightly joined; that they shall be so firmly supported that they will not bend or break by their own weight, or through the changes of temperature to which they are subject, and that they shall be, if not always in plain sight, at most only hidden by some covering easily removed, are points which the commonest kind of common sense would not fail to observe."

CHAPTER XI.

PLUMBING FIXTURES.

Description of Plumbing Fixtures.

THE selection of proper plumbing fixtures will next require our attention. To use cheaper and inferior articles would be, in all cases, a mistaken economy. Such cheap fixtures not only wear out much sooner than a good, although more costly, article of manufacture, but they will often require repairing, patching up, and the bills for the latter work will, in many cases, exceed the amount of the first expenditure. A wise householder will reduce the number of his plumbing fixtures to a minimum, but will choose none but first-class appliances.

How to arrange such fixtures properly, where to place them, and how to keep them sweet and clean after use, will be discussed further on. We will first briefly review the kind of fixtures to be used, without, however, giving a detailed description of all appliances at present in the market.

The *kitchen sink* may be of cast-iron, and can be painted, galvanized or enamelled, or it may be of soapstone. Galvanized or enamelled sinks will last only a few years, after which the galvanizing wears out, while the enamel scales off. Plain painted sinks require frequent renewal of the paint, and cannot be recommended. Soapstone sinks are better, but they soon assume a dark color and a greasy appearance. Sinks of wrought-steel have lately been introduced into the market, and may prove to be economical and not easily worn out if protected against corrosion by the Bower-Barff rustless process. The neatest, most cleanly and best sinks for use are those in earthenware, which are imported from England, and are made in all required sizes. Their external beauty—especially if put up in a light open frame, with a marble back, and set on a tiled floor—and the fact that they are non-absorbent, will soon make earthen or porcelain sinks the general favorites amongst house-wives.

Kitchen sinks, of whatever description, should have the outlet protected by a fixed strainer to prevent obstructions of the waste pipe. The waste pipe, generally a 1½ inch pipe, which size is ample in all cases, should be trapped directly underneath the sink by an efficient trap. As the kitchen sink is in constant use, there is no danger in using an S-trap, protected by a vent pipe against siphonage. The connection of the vent pipe should be at the crown of the trap, and to prevent stoppage of the vent pipe at the crown by soap-suds or grease, it should preferably be made as shown in sketch, Fig. 207. Lead or brass traps might, with advantage, be cast with a funnel-shaped vent pipe attachment, such as shown in the sketch. As kitchen sinks are generally located on the basement floor, an overflow pipe is unnecessary. For sinks in small households,

I should advise against the use of any grease trap, as the grease may, with advantage, be saved and not poured out into the sink. For large mansions, restaurant kitchens, boarding houses, the

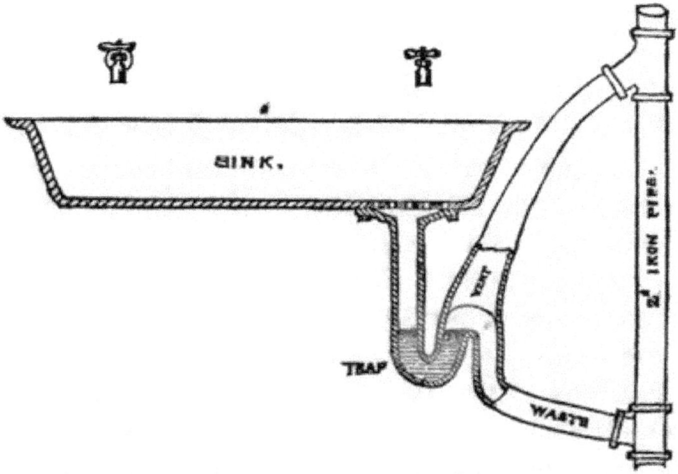

FIG. 207.—Kitchen Sink trapped by a vented S-trap.

use of a grease trap, preferably one located outside of the house, is recommended. Grease traps inside a dwelling, under or near a pantry or kitchen sink, will, in most cases, become a cesspool, and prove a serious nuisance, unless carefully and often cleaned.

The neatest material for *laundry tubs*—and, in fact, for nearly every fixture in the house—is porcelain, as it is non-absorbent and most cleanly. Such tubs are, of course, more expensive, and for this reason are not generally used. Soapstone tubs, as well as cement stone tubs, answer very well, the latter being made in one piece without seams.

The waste pipe for a set of three or four tubs should be 1½ inches diameter; an overflow pipe is not necessary, except where the laundry is located on the top floor of the house. Theoretically, each tub should have a separate trap, but in practice, one trap is almost always used for the whole set, placed either at one end of the set or under the middle tub. If the laundry tubs are in constant use, a properly vented S-trap or a running siphon trap of the Du Bois pattern should be used; if not in constant use, it may be advisable to use one of the mechanical traps.

Pantry sinks are generally made of copper, either with oval-shaped bottom or with a flat bottom. Small earthenware sinks, for the butler's pantry, are very clean and attractive in appearance, but with them glass and crockery ware are more exposed to frequent breakage. Copper sinks should be tinned and planished, and the copper must have a weight of not less than 18 oz., better 24 oz. per sq. foot. They are frequently closed by a plated plug or stopper, but sometimes the waste pipe is closed by a waste cock. In both cases an overflow pipe is used, connected to the waste pipe below the water seal of the trap. Better than either arrangement would be a short standing overflow, inserted into the socket of the waste pipe (see Fig. 208), thus doing away with a possible nuisance, the separate overflow pipe. In order to prevent the standing waste from being in the way while washing dishes, we would suggest to have the pantry sink flat-bottomed, with a slight slope towards the outlet, and made with a recess for the standing waste, as shown in the drawing. The waste pipe for the sink should be not larger than 1½ inches ; 1¼ inch is preferable, and the trap

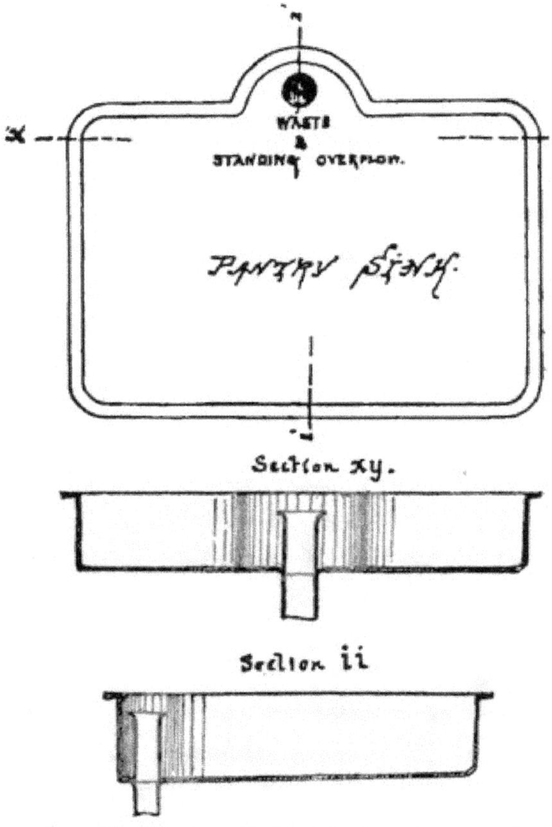

FIG. 208.—Pantry Sink with standing overflow.

should be of this same size. Pantry sinks for hotels or large establishments, should have a grease trap to intercept the fat removed by washing dishes and plates.

Refrigerators should not have waste pipes connected directly with any drain or sewer. Those of smaller size may waste into pails, removable by hand. Larger sizes should waste into an open cup or tray, connected to a drain or soil pipe by a trapped waste pipe, provided with a tight-shutting stop-cock on the line of the waste pipe, which should be shut each time the refrigerator is put out of use.

Stationary wash-stands have been extensively used, not only in bath-rooms or lavatories, but in sleeping-rooms, in nurseries, in hospital wards, and in offices. Being in almost every instance defective in construction and general arrangement, they have, in innumerous instances, been the cause of sufferings, headache, general debility, sickness or fatal illness. Since the public has become aware of the dangers and risks connected with cheap or dishonest plumbing work, a great cry has been raised against stationary wash-bowls, and most of the sickness due to imperfect drainage in general has been attributed to this one fixture. This view, however, is erroneous, for *any kind* of plumbing appliance not properly trapped and ventilated is a danger and risk to health, if placed in a living or sleeping-room or office, and, on the other hand, there is no good reason why a wash-bowl might not be fitted up in as perfect a manner as a water-closet, a sink or a bath-tub, if placed in a well-ventilated bath or dressing-room.

Wash-basins should be of porcelain, and are made either round or oval in shape. (See Fig. 209a and 209b). Oval wash-bowls have been introduced only recently, and are appreciated by many as being of a more convenient shape. As commonly fitted up, wash-bowls have on their bottom a socket and coupling, to which the waste pipe is attached. The bowl is closed so as to hold water

FIG. 209a.—Round Wash-bowl.

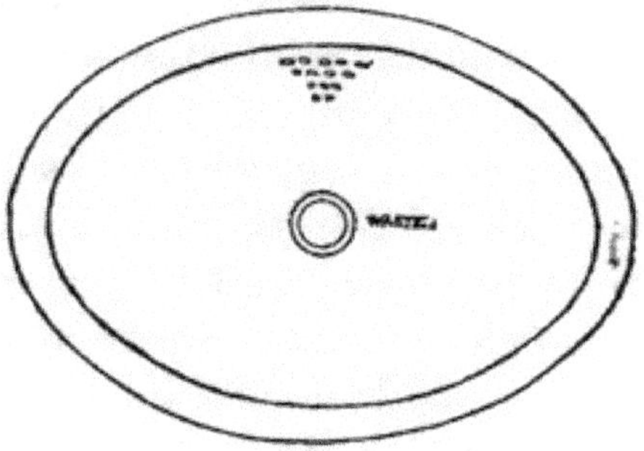

FIG. 209b.—Oval Wash-bowl.

by means of a plug, to be inserted into the socket. The bowl requires in this case an overflow pipe, which, as I have repeatedly stated, remains imperfectly flushed, and is often ill-smelling.

To insure cleanliness of the bowl, it should be possible to empty it quickly. A novel bowl, which is said to be quick-emptying, has lately been manufactured in England, and is shown in vertical section in Fig. 210. The outlet is supposed to be closed by a waste-valve at some distance from the bowl.

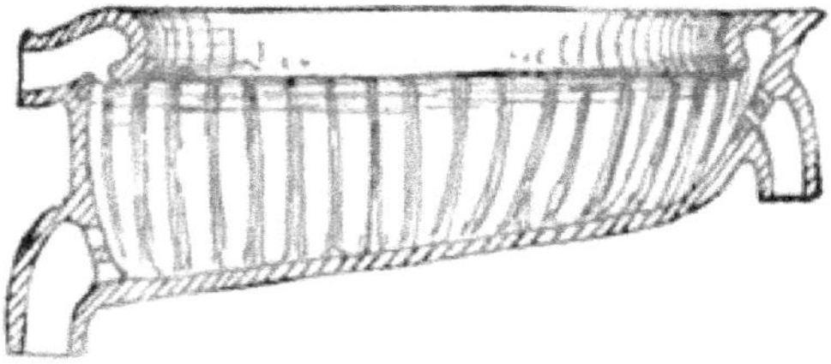

FIG. 210.—Tylor's Quick-emptying Bowl.

This bowl, as well as others of the common round shape, are now made with a flushing rim on top, the hot and cold water being introduced by a nozzle and entering the bowl simultaneously at all sides, giving it a thorough cleansing. This is certainly a great improvement, which recommends itself at once for lavatories in hotels, hospitals, barracks and club-houses. (Fig. 211).

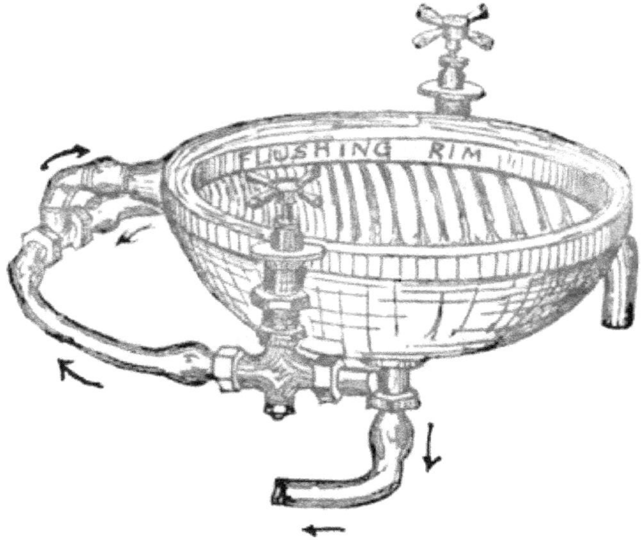

FIG. 211.—Flushing Rim Wash-bowl.

If not closed at the bottom by a plug, wash-bowls are fitted up with waste-valves, which in most cases take the place of the objectionable overflow pipe, while in a few of them the ordinary overflow is used.

Fig. 212 represents the top of a wash-bowl,

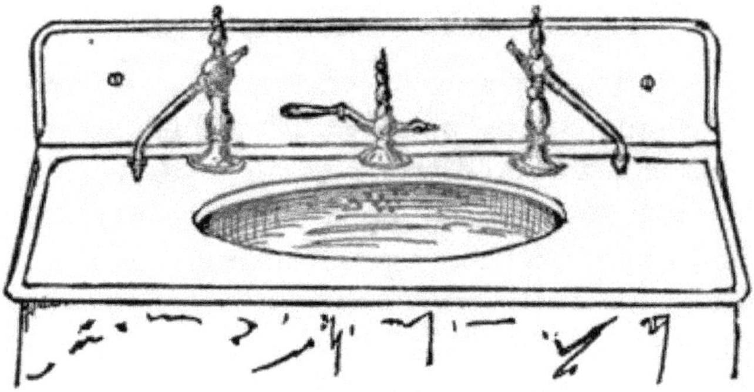

FIG. 212.—Boston Waste-Cock.

showing the supply cocks, and in place of the usual chain stay, chain and plug, the handle of a Boston waste, which is simply a ground cock with large water way on the waste pipe. The overflow pipe must join the waste pipe beyond the waste-cock.

Fig. 213 shows Foley's waste valve for basins, which dispenses with a special overflow pipe, as the overflow takes place through the hollow center of the waste valve.

Boyle's waste valve (Fig. 214) is very similar in principle to the preceding one, but here the waste and overflow are shaped in one piece with the bowl, and the hollow stand pipe, forming the valve, and overflow pipe, is inserted into the space at the rear of the bowl.

FIG. 213.—Foley's Waste Valve.

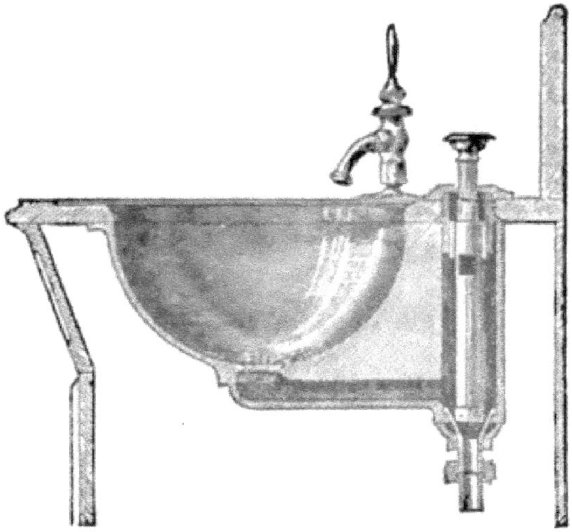

FIG. 214.—Boyle's Waste Valve.

Bennor's secret overflow basin differs from the above merely in having the handle which lifts the valve coming out below the slab in the bowl, not on top of the slab. (See Fig. 215).

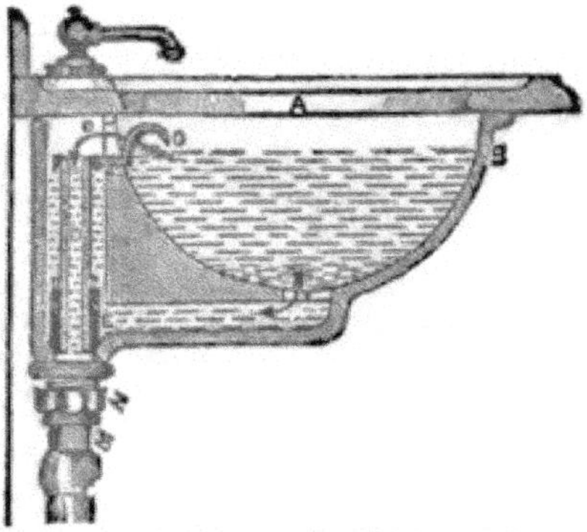

FIG. 215.—Bennor's Waste Valve.

McFarland's waste (Fig. 216) for basins requires a short overflow, and has a valve, shutting off the

FIG. 216.—McFarland's Waste.

waste, and having a number of holes in its body to allow the water from the overflow pipe to pass through, should such an overflow occur.

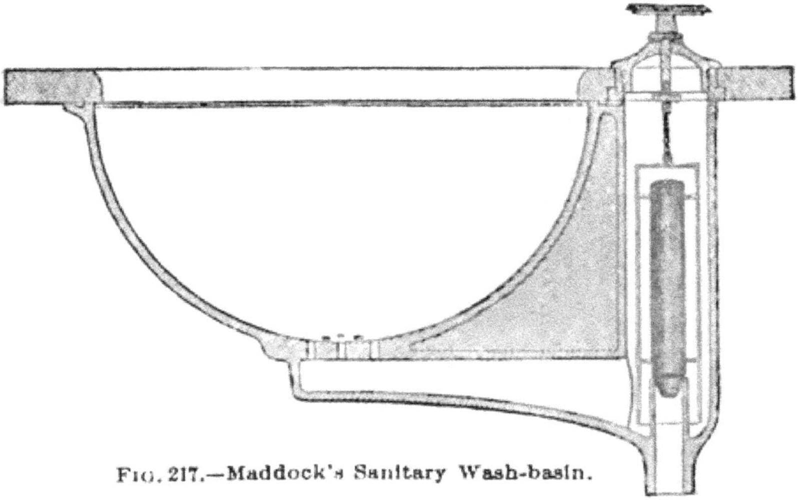

FIG. 217.—Maddock's Sanitary Wash-basin.

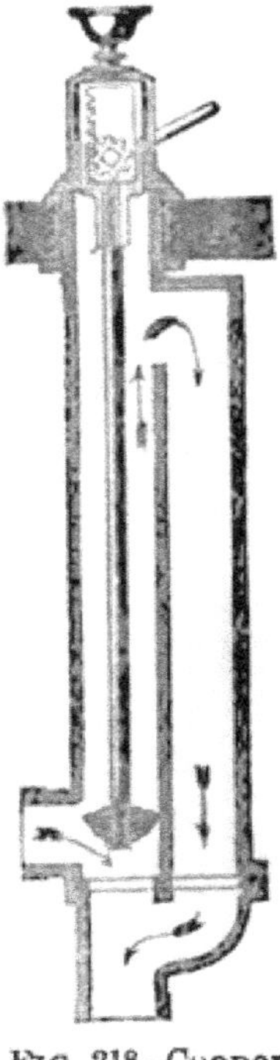

FIG. 218.-Cooper, Jones & Cadbury's Secret Overflow Valve.

Maddock's sanitary wash-basin (see Fig. 217) is somewhat similar to Boyle's valve, and its construction can be easily understood from the sketch.

Cooper, Jones & Cadbury's secret overflow valve is illustrated in Fig. 218. The cut shows plainly the manner of lifting the plug or valve to discharge the bowl, and likewise the course that any overflowing water would take, if the valve should be closed.

Moore's no-overflow basin valve (Fig. 219) differs from those mentioned, in dispensing altogether with the overflow pipe. This result is attained by attaching to the rod of the waste valve a large copper float. If the water rises in the bowl to the height shown in the cut, the

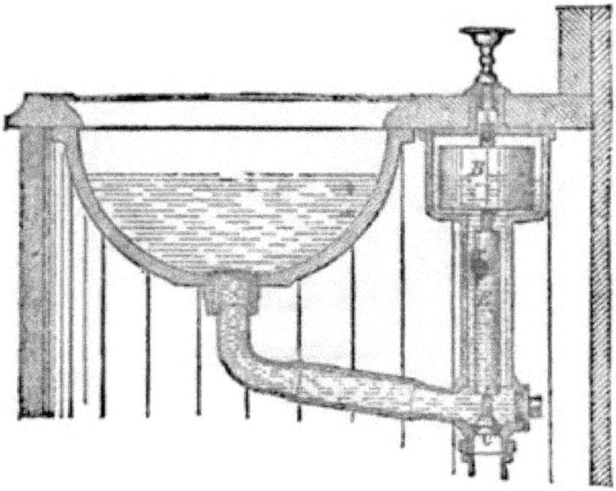

FIG. 219.—Moore's No-overflow Wash Bowl

float commences to lift the valve, and the water will run out of the waste pipe until its level is so

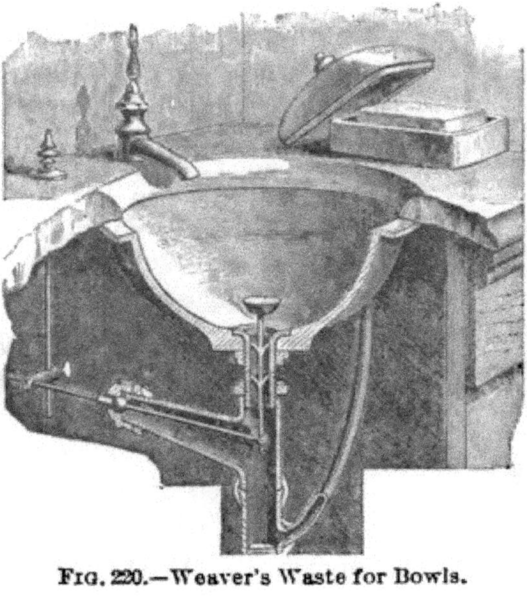

FIG. 220.—Weaver's Waste for Bowls.

much lowered as to allow the float and, consequently, the valve to drop again in its seat.

Fig. 220 illustrates a wash-basin fitted up with Weaver's waste. As will be seen, this shut-off differs from all previous ones, by closing the basin directly at its outlet, in the place where the usual socket and plug is arranged. By pushing the knob, the plug is raised by means of the lever shown. The overflow pipe joins the waste pipe below the Weaver waste.

Quite similar to Weaver's waste is Stidder's waste valve for basins, shown in Fig. 221, an

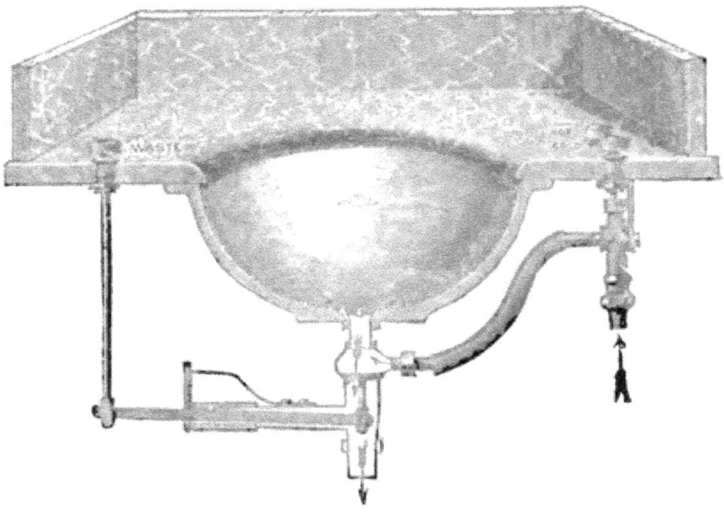

FIG. 221.—Stidder's Waste for Bowls.

English arrangement. The supply is shown to enter the bowl at the bottom, just above the plug, which closes the outlet. Such a method of supply is not usual in American plumbing.

A large number of other waste-valves are

in the market, more or less similar in principle and construction to the ones described above.

There can scarcely be any doubt about the convenience of such waste valves as compared with the chain and plug arrangement. With few exceptions, however, these waste valves close the basin at a great distance from the outlet strainer, and foul matter, left from previous use, may mingle with the clean water drawn, which is anything but agreeable. Some of the plugs will close imperfectly if hair or lint catches at the seat, and in such a case it is impossible to hold water in the bowl, to the great annoyance of the person intending to use it. Although the valve chambers are mostly accessible for cleaning purposes, the latter operation is easily neglected and foul slime may accumulate in the valve chamber, the putrefaction of which would soon be the cause of annoying odors.

Tip-up lavatories (Fig. 222) have been in use

FIG. 222.—Tip-up Basin.

many years, and nobody can doubt their great convenience. They do away with the objectionable chain and plug, and, at the same time, dispense with the overflow pipes altogether.

The only criticism that can be raised against them is the possibility of the lower bowl becoming foul. This objection has been overcome recently by arranging the upper bowl so as to be easily lifted and removed, and by making the lower bowl or receiver, quick-emptying. Even then the cleanliness of such apparatus depends, of course, upon the care and conscientiousness of the servants.

The author would suggest a self-cleansing wash-bowl, without overflow pipe, as follows:

A tip-up basin, the top bowl to be arranged so as to be easily lifted for cleansing purposes. The lower bowl to be arranged with outlet near back so as to be quick-emptying. The top of the lower bowl to be provided with a flushing rim, and a supply pipe to this flushing rim so arranged by means of stop-cocks, levers and valves, that each time the upper bowl, which is hung on pivots, is tilted, the supply valve is opened, allowing a liberal flush of water to rush through the flushing rim into the lower basin.

The waste pipe for wash-bowls need, in most cases, not be larger than 1¼ inches diameter, except where the pressure of water in the supply pipes is very large, furnishing a heavy stream of water. In this case the waste pipe must be made large enough to remove all the water without danger of an overflow.

If the bowl is in constant use, an S-trap, properly vented (see Fig. 207), is the best trap for its waste pipe. If located in bath-rooms, which are not always in use, a different method of trapping may be advisable. No general rules can be stated applicable to all cases; we refer, however, to the chapters on traps, especially Chapter VI., page 121 to 130.

The best and neatest non-absorbent *bath-tubs* are those of porcelain, imported from England. Their great cost, and the heavy weight of each tub, offer obstacles to their more general introduction in other dwellings than those of the wealthy. Their use in hospitals and public bathing-houses should be much encouraged. In this country, the majority of bath-tubs are made by lining a wooden box, blocked out to desired shape, with tinned and planished copper, weighing from 10 to 24 ozs. per sq. ft. Such tubs are very good, except that they require, for appearance's sake, a casing of woodwork, which, for sanitary reasons, should be done away with as much as possible.

Iron bath-tubs, lined with a porcelain enamel, are made to stand on legs free on the floor. They are a very satisfactory article, the only objection to their use being the scaling off of the enamel. In Europe, bath-tubs are made of metal sufficiently heavy to stand without wooden framing or lining. Sometimes tubs are made of slate or marble slabs, cemented together; in this case it is preferable, for convenience's sake, to lower the tub one or more steps, which arrangement is often adopted for public bathing establishments.

The usual bath-tubs in American dwellings have wastes closed, at or near the bottom, by a plug, with chain (Fig. 223), or by a waste cock of some kind, in

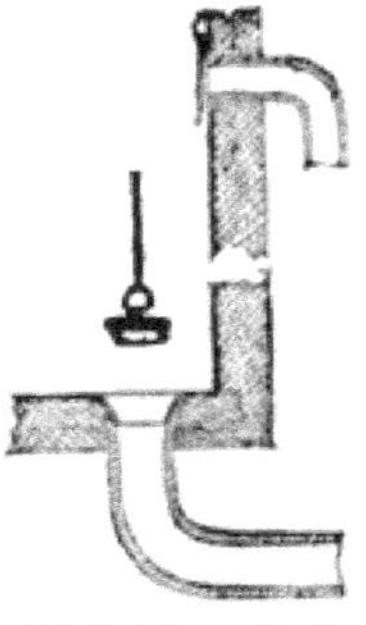

FIG. 223.—Common Chain and Plug.

which case a common overflow pipe or a channel in the waste valve for such overflow—both objectionable for well-known reasons—are required. Of waste valves for bath-tubs, there are a large number, but only a few of these will be illustrated.

Fig. 224 shows the Boston waste valve, which consists of a ground cock, with large water way to allow a quick emptying of the tub. It is similar in construction to the Boston valve for basins.

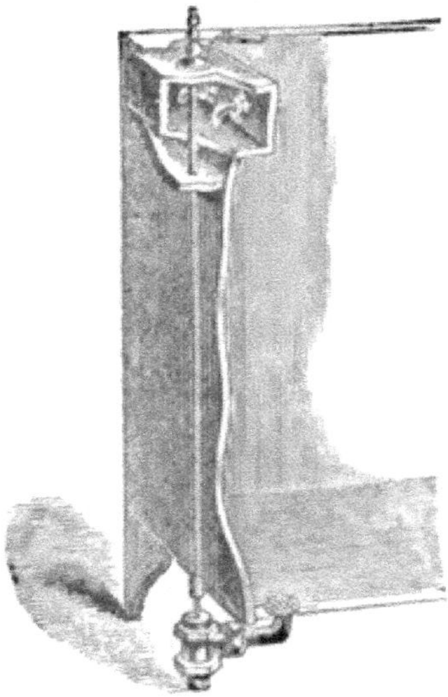

FIG. 224.—Boston Waste Cock.

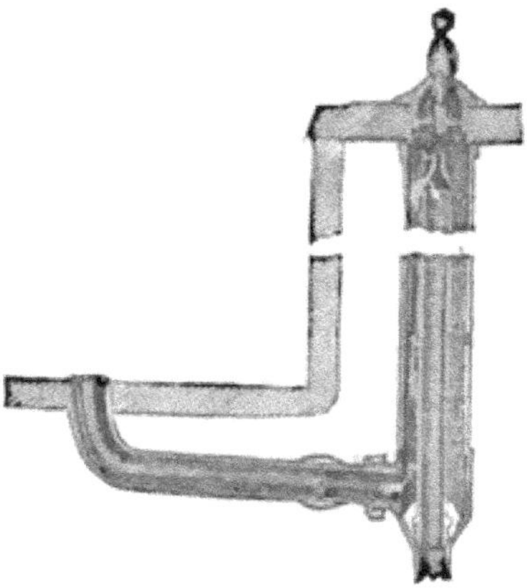

FIG. 225.—Foley's Waste Valve.

Fig. 225 illustrates a bath-tub, fitted up with Foley's waste valve, which dispenses with a special overflow pipe.

The same is true of McFarland's waste, which is shown in Fig. 226. Weaver's waste for bath-tubs (see Fig. 227) is similar in construction to his basin waste, and needs no further explanation.

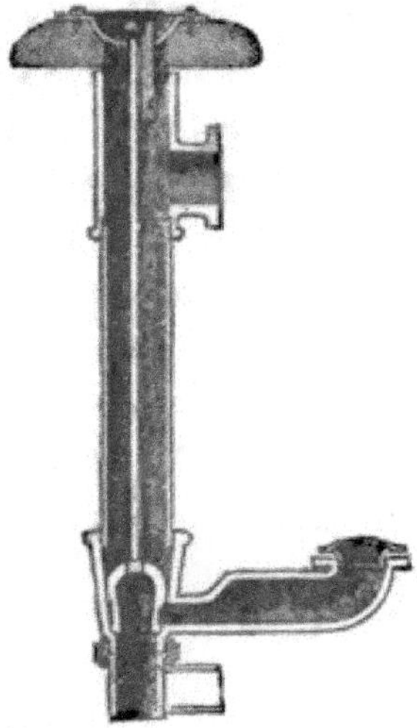

FIG. 226. — McFarland's Waste.

FIG. 227-Weaver's Waste for Bath-Tubs.

The comments made above in regard to basin wastes, are also more or less true of these wastes.

FIG. 228-Standing Overflow.

I decidedly prefer the arrangement known as a standing waste, and used extensively in Boston. (See Fig. 228.) It does away with the chain and plug, the waste cock and the overflow pipe, and is most cleanly and simple in operation.

The bath-tub, if copper-lined, can easily be arranged with a recess at the foot of the tub for the standing waste, similar to the one proposed in Fig. 208, for pantry sinks.

As regards the trapping of the bath waste, the remarks made for wash-bowls have reference to bath-tubs as well.

House-maids' sinks are neatest if made of earthenware; the same is true of *slop sinks*, which are often fitted up in dwellings to empty chamber slops. The latter kind of sink is generally much deeper than the ordinary sink, or else it is shaped like a hopper, with a flat square sink on top, provided with a flushing rim. (Fig. 229.) The less

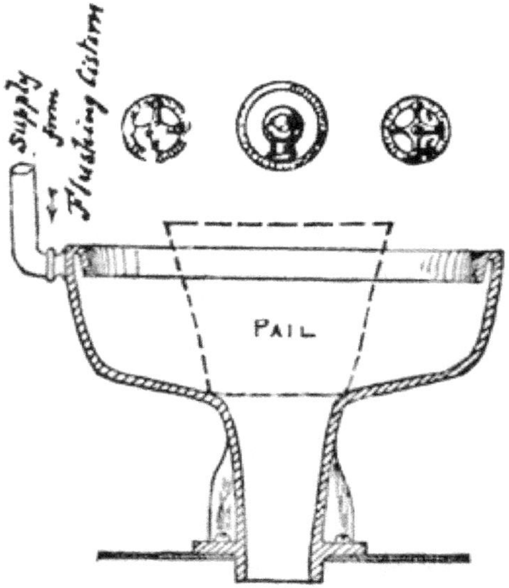

FIG. 229.—Flushing Rim Slop-hopper in Section.

surface of such hopper or sink is exposed to fouling the better, therefore the top should be kept of moderate size and not unnecessarily enlarged. It is of the utmost importance that these sinks and slop-hoppers should be flushed out after use, and to do this efficiently I always recommend to fix over the sink, or hopper, a small water cistern, either a valve or else a siphon cistern, operated in each case by a chain and pull. The outlet of such sinks, or hoppers, should be protected by strong strainers or baskets to prevent the throwing in of objectionable solid articles, such as brushes, rags, etc., which would probably remain in the trap or choke the waste pipe. The trap used, if an S-trap, must be efficiently protected by a very large branch air pipe taken from the crown of the trap, for, if a pail of slops is suddenly poured into a slop-hopper, the trap would otherwise lose its water seal by siphonage.

Urinals should not be fitted up in private houses. It is a most difficult matter to keep them clean and neat. Water-closets should be fixed with as little wood work as possible (see Figs. 251 to 255), and then they may be used as urinals. The latter are necessary only in public buildings, public places, and in office buildings. The neatest fixture for such use is the porcelain-lipped "Bedfordshire" urinal.

There is now for sale a pattern in which the basin is so shaped as to hold a certain quantity of water. (See Fig. 230.) This type of urinal is preferable to the old-fashioned style, as an immediate dilution of the urine takes place. Urinals in office or public buildings should always be flushed from a special tank, which may be either a siphon or valve cistern, operated by hand, or by treadle or door action, or it may be one of the many automatic flush tanks, (See Fig. 230a and b). To flush the urinals from

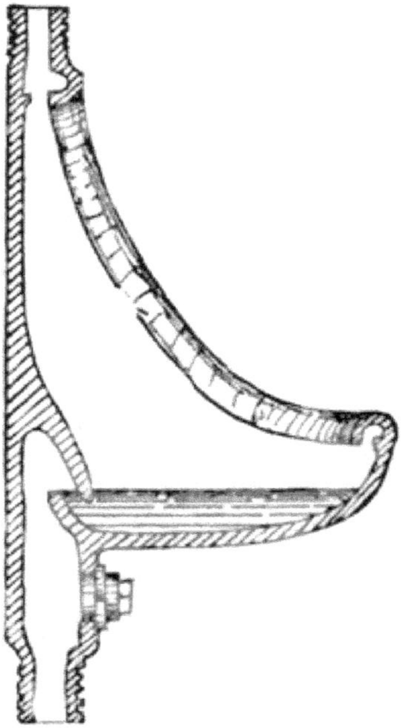

FIG. 230.—Urinal, with Basin to hold water.

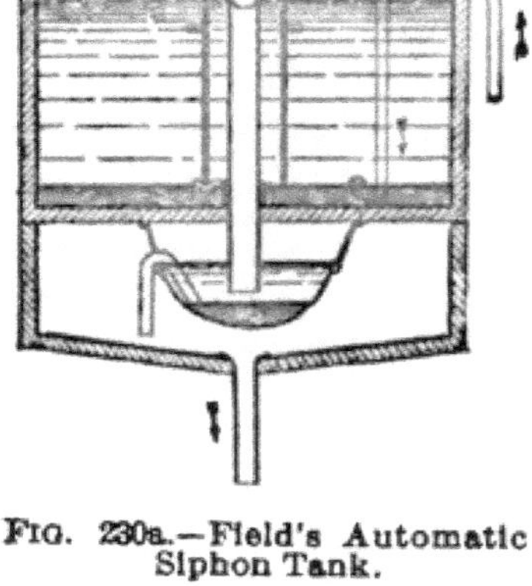

FIG. 230a.—Field's Automatic Siphon Tank.

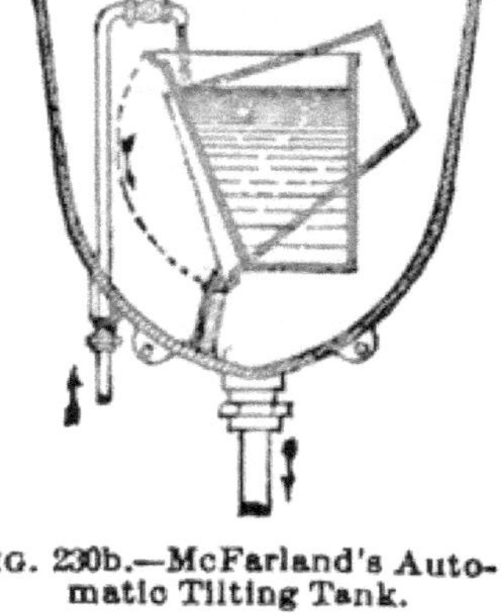

FIG. 230b.—McFarland's Automatic Tilting Tank.

a small branch pipe and a self-closing bibb, to be opened by the person using the urinal, is objectionable in all cases, and absolutely inadmissible in the best kind of work. A flushing tank, operated by chain and pull, may be used for toilet-rooms of private offices, where intelligent attention to the required flushing may be expected; in all other cases, an automatic supply is preferable.

The waste from a single urinal need not be larger than 1½ inches; for a row of urinals a 2″ pipe may be required. The trap should, in any case, be as small as possible—about 1¼″—so as to have its contents thoroughly changed at each flush. Where urinals are flushed automatically and are in constant use, an S-trap, properly vented, can be used without danger.

We must lastly consider the most important plumbing fixture of the dwelling—*the water-closet.* A proper and satisfactory selection of this fixture is rendered very difficult on account of the large number of water-closets now for sale. Wherever advice is sought by prudent householders on "sanitary drainage," no question is probably put as often as this: *Which is the best water-closet?* or *What water-closet would you recommend me to use?*

Without going into a detailed description of the various water-closets in use, the author will endeavor briefly to answer this question. Generally speaking, water-closets may be divided into two distinct classes:

1, those *with mechanical parts* or movable machinery—the pan-closet, the valve-closet, and the plunger-closet.

2, those *without any movable machinery*—the hopper closet and the washout-closet.*

The pan-closet (Figs. 2 and 4) was described and condemned in Chapter II., page 13 ; the valve-closet and the plunger-closet were illustrated in Fig. 12 and Fig. 13, and their defects pointed out.

The author's experience has led him to advise in every case against closets with movable parts, as being complicated, easily deranged, and readily fouled. While admitting that some of the closets of this description are of first-class make, the writer has never considered them fit for use for sanitary reasons.

Closets of the second class only should be used in sanitary homes. With these, all the machinery is located in the flushing cistern, fixed at a proper height above the water-closet bowl. The water-closet itself is merely a plain bowl—of earthenware in first-class dwellings—with a flushing rim on top of the bowl, to which the water is brought by a supply pipe of large diameter.

Hopper-closets may be subdivided in *long* and *short hoppers*. The former have a trap (of iron or lead) below the floor (Fig. 231), while the short

---

* For a detailed description of water-closets, the reader is referred to Prof. T. M. Clark's Articles on "Modern Plumbing" in the *American Architect*; Prof. T. M. Clark's Book on "Building Superintendence;" Mr. Glenn Brown's Articles on "Water-Closets," in the *American Architect*; Papers on "Sanitary Plumbing," in the *American Architect*; W. P. Gerhard's Paper on "House Drainage and Sanitary Plumbing" in the Fourth Annual Report of the R. I. State Board of Health; W. P. Gerhard, House Drainage and Sanitary Plumbing, 2d edition, published by D. Van Nostrand, 1884.

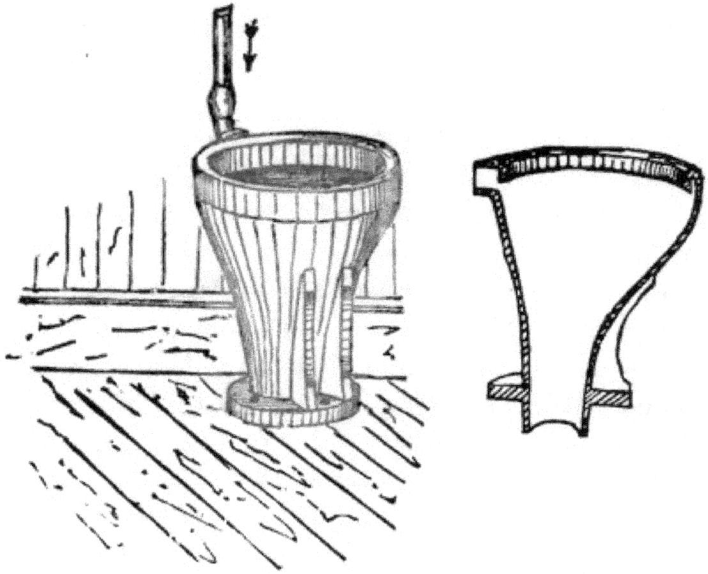

A B

FIG. 231.—Long Flushing Rim Hopper.

hoppers have the trap (of iron or earthenware) above the floor (Fig. 232). The latter are pre-

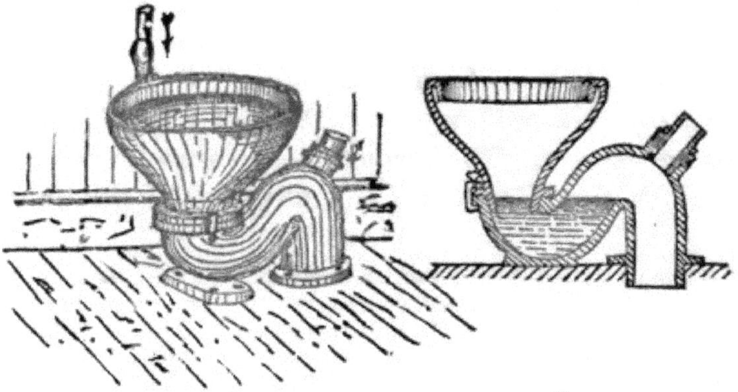

A B

FIG. 232.—Short Flushing Rim Hopper.

ferable, for the surface exposed to fouling is much smaller than with long hoppers; the trap is in sight, which is a great advantage; finally the level of the water (in the trap) is nearer to the seat.

The *ideal* water-closet, however, has yet to be invented, and—as in so many other matters—we must content ourselves with the best approximation to the perfect apparatus.

Unless fitted up with skillful judgment, hopper-closets, both short and long, are apt to cause dissatisfaction through an occasional fouling of the bowl. Without a properly arranged system of flushing, they are apt to become extremely nasty and foul, especially so in water-closet apartments of public buildings of any kind. To prevent this, the water supply should be ample, the amount for each flush should be discharged rapidly through a large supply pipe into the flushing rim, and the precaution must be observed to give the closet at each use a preliminary wash, to moisten the sides of the bowl—and an after-wash, to thoroughly rinse the closet and expel the soil from the trap.

To secure to such plain closets without mechanical apparatus, the advantages which the valve and plunger-closets have of holding a large surface of water in the bowl, a modified form was adopted, which the writer has called a *washout-closet*, while other writers call it an improved hopper-closet.

Washout-closets may be subdivided into:

1, those having a bowl shaped so as to hold water (generally not more than 1½ inches in depth) and having under the bowl, generally in one piece with it, a siphon trap (Fig. 233).

2, those in which the basin itself is shaped so as to make a water-seal trap, and holds water to a

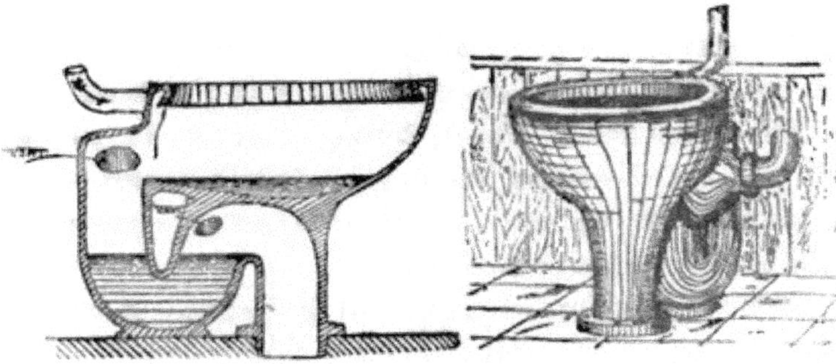

A B

FIG. 233.—Washout-Closet, with water in basin and trap below basin.

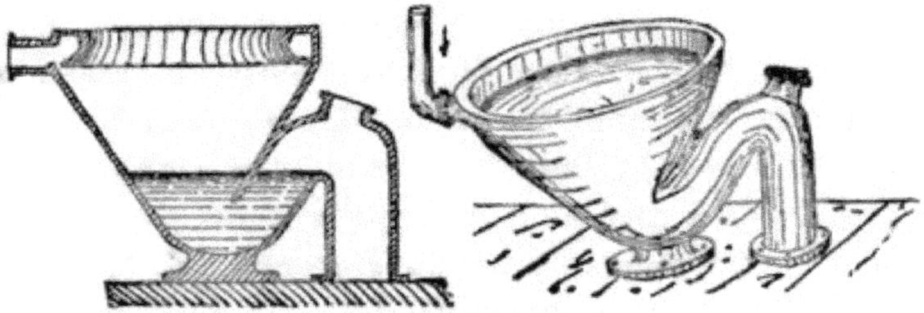

A B

FIG. 234.—Washout-Closet, with basin holding water, and shaped so as to form a seal.

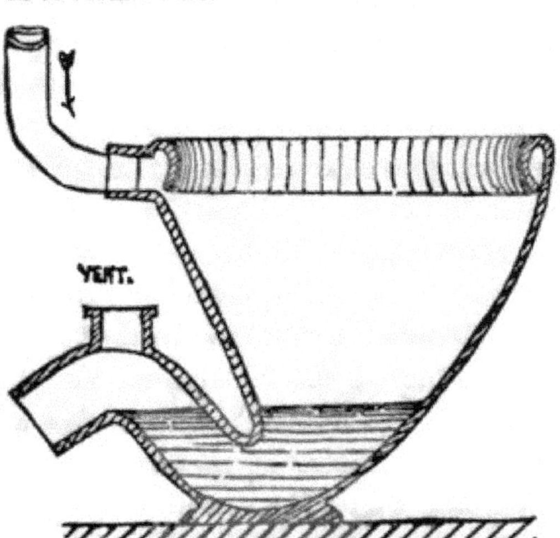

FIG. 235.—Author's Suggestion for Improved Short Hopper-Closet.

greater depth than those first mentioned (Figs. 234 and 235).

Against the first-mentioned closet the criticism may be raised that the force of the flush is largely exhausted in cleansing the basin, after which the water, soil and paper, drop into the trap, which is not exposed to view as with other closets.

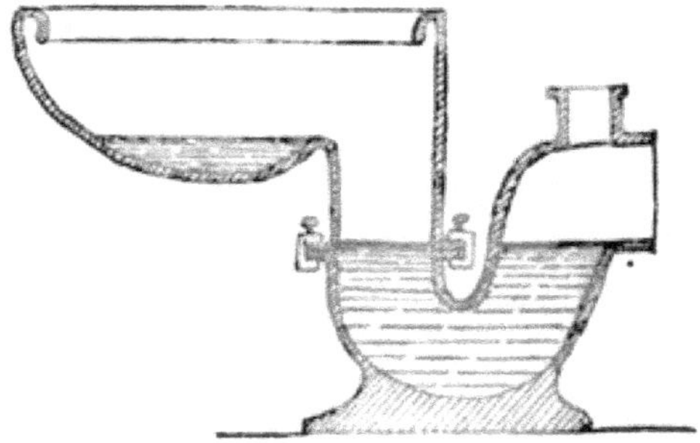

A

B

FIGS. 236a and b.—Author's Suggestion for Improving Washout Closet.

There is, consequently, chance of foul matter lodging in the trap, to give off offensive gases. There is also some danger of foul matter accumulating in the vertical shaft between the bowl and the trap. The evil may, perhaps, be remedied by shaping the top of the bowl and its flushing rim as indicated in Figs. 236a and b.

The other closets, in which the basin itself forms a trap against gases, are preferable. They much resemble in principle the short hopper, having the advantage of being made in one piece of earthenware, of holding more depth of water in the bowl, and of having a larger surface of water. These closets, in the writer's opinion, approach the ideal closet more closely than any other kind. One difficulty has yet been but partially overcome with most of them, namely, the proper flushing and cleansing of the basin, especially the proper discharge of soil and paper. Ingenious siphon arrangements and water jets have been invented to effect this purpose. It is quite possible, however, that a simple discharge of water from a cistern through a good flushing rim or a series of fan-washers, arranged on the top of the bowl, will suffice to expel all matters from the basin, if the latter is of proper size and shape. (See Fig. 235.)

The illustrations (Figs. 2, 12, 13, 231, 232, 233 and 234) show all the *types* of water-closets now in the market. There is a striking contrast, as regards simplicity of construction, between the pan, valve and plunger-closets on the one side, and the hopper and washout-closets on the other. Any of the closets of the latter class, if bought from a responsible, first-class manufacturing firm, is likely to give satisfaction, provided the closet is well taken care of, for even the best kind will need repeated cleaning, washing and scrubbing. The latter operations will be much facilitated by a proper arrangement of the closet, and here again the plain earthen hoppers, and washout-closets, are vastly superior to mechanical closets.

It is interesting to notice what the Report of the General Board of Health of England, made in 1852, says about the principles of the construction of the water-closets.

"It is now necessary to revert to the construction of the chief apparatus for the decent and efficient sanitary arrangement of every household, an apparatus to which but little attention is usually paid, but which requires the most serious consideration, as one of the primary works for the sanitary improvement of houses and towns, namely, the water-closet.

The particular points to be sought for in the construction of the apparatus in question, appear to be:

1. A scour for the complete removal of the soil.
2. The best trap against the ingress or regurgitation of effluvia from the general system of drainage and sewerage with which each soil-pan or house-sink must communicate.
3. The consumption of the least quantity of water for a complete scour and perfect trap.
4. Durability, or freedom from the liability of—
   a. Breakage in consequence of frost.
   b. Derangement of the machinery.
   c. Breakage by careless usage.
   d. Stoppages.
5. Easy repair.
6. Cheapness when manufactured on a large scale."

The report, further on, condemns the pan-closet, and recommends simple hopper-closets, preferably short hoppers, and a modified form of hopper, such as shown in Figs. 237 and 238. It is a somewhat remarkable fact, that notwithstanding such a severe but just condemnation of the pan-closet more than thirty years ago, it should still be found in most dwelling-houses of to-day.

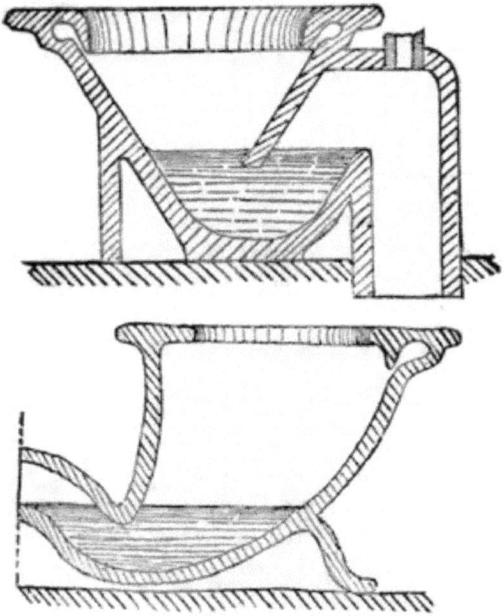

FIGS. 237 and 238.—Closet Forms recommended in 1852 by the General Board of Health of England.

General Arrangement and Care of Fixtures.

Having described the quality and character of fixtures to be used, we must next speak about their general arrangement. We have repeatedly stated that it is highly desirable to have everything relating to plumbing in plain sight. Traps concealed under floors should be abolished ; soil and vent pipes buried in walls or partitions, fixtures encased in tight carpentry, and supply pipes with stop-cocks that cannot be immediately reached when necessary, are highly objectionable.

There is a certain prejudice against having plumbing appliances left without any casing or covering, especially on the part of women, but we must gradually educate our good house-wives in these matters, and we venture to say that if the objections against the old methods would be properly explained to them, very few only would object to the advice of sanitarians, to have every fixture open and accessible. If all women would be as practical and exhibit as much good sense as Jill in Mr. E. C. Gardner's charming book, "The House that Jill Built", sanitary inspections would soon be rendered unnecessary, and the annual plumbers' bills for repairs would become a thing of the past.

"I wish it were possible," said she, "to build a house with everything in plain sight, the chimneys, the hot-air pipes from the furnace, if there are any, the steam pipes, the ventilators, the gas pipes, the water pipes, the speaking tubes, the cranks and wires for the bells—whatever really belongs to the building. They might all be decorated if that would make them more interesting, but even if they were quite unadorned they ought not to be ugly. If we could see them we shouldn't feel that we are surrounded by hidden mysteries liable at any time to explode or break loose upon us unawares. Those things that get out of order easily ought surely to be accessible. I don't believe there would have been half the trouble with plumbing, either in the way of danger to health or from dishonest and ignorant work, if it had not been the custom to keep it as much as possible out of sight. There is a great satisfaction, too, in knowing that everything is genuine."

The following advice of a physician in "The House and its Surroundings," is equally well to the point:

"As to the pipes above the basement, you should insist upon having them all, within as well as without the house, as accessible as possible. Plumbers, as the late Dr. Parkes remarks, 'try to conceal everything,' and, in consequence of this principle, when any accident occurs, the house is pulled about and the walls and woodwork damaged to a great extent, because no one knows or can get at the exact direction of the offending pipe. Therefore, all these pipes, including their inlets and outlets, should be visible, or, if enclosed at all, should be cased in with wooden coverings, lightly screwed together, and not, as is usually the case, imbedded in plaster or cement, or otherwise fixed securely into the main or other walls of the building."

The following quotation from the well-known English architect, Ernest Turner, referring to service pipes, might be applied in general to plumbing-work. He says:

"Service pipes are commonly kept carefully out of sight. This is an excellent arrangement—for the plumber—who is thus enabled to conceal any amount of scamped work. For the owner, its drawbacks are three, at least.

1. It makes defects or accidents more difficult of detection.
2. It makes them more mischievous in action.
3. It makes them more costly in correction.

No pipes above ground, as was said in the preceding chapter, should be hidden behind anything but a *hinged* casing."

There should be as little as possible wood-work around plumbing fixtures, and this will not at all detract from the appearance of such work, provided the work itself is properly done and well finished. If the space under and around bowls, sinks, tubs and water-closets is kept entirely open, cleaning operations are much facilitated, and everything is readily inspected at any time, without the necessity of using tools to remove boards or casings. An open arrangement of fixtures is equally well adapted to offices, small dwellings, or the most luxurious residences. Fancy and orna-

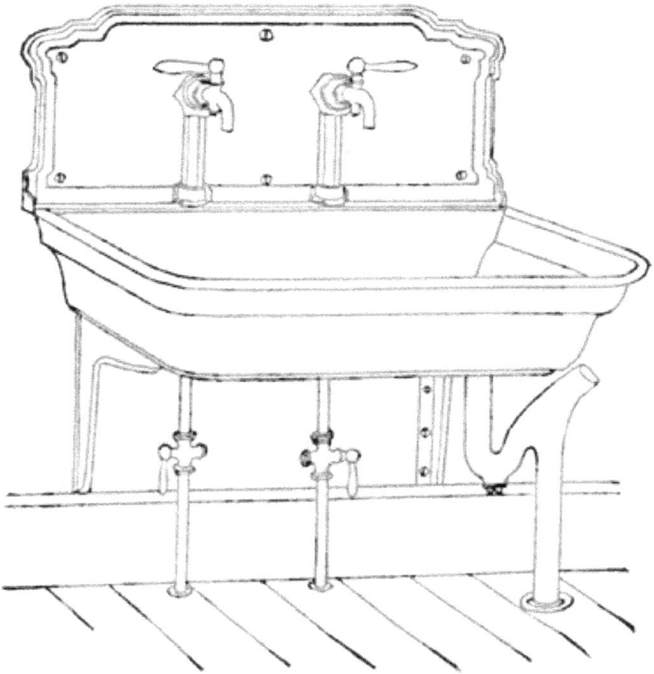

FIG. 239.—Iron Kitchen Sink, supported on brackets.

mental casings of woodwork have hitherto been considered indispensable for finishing bath-rooms. The mistaken notion of judging the quality of a job of plumbing by the costliness of the marble slabs, the silver-plating of the faucets, the decorating and gilding of basin and water-closet bowls, the expensive hard wood finish, has gradually and slowly given way to a better appreciation for fixtures properly trapped, amply ventilated, and well flushed.

The following remarks and sketches are chiefly intended to explain a proper and sanitary method of fitting up modern conveniences :

FIG. 240.—Earthenware Kitchen Sink, supported in a frame, resting on decorated legs.

Kitchen sinks may be supported on brackets, securely fastened into the walls, or else they may rest on legs on the floor. Fig. 239 shows the former arrangement for an iron sink, with iron back, while Fig. 240 shows the latter, for an earthenware or "Imperial" sink, the illustration being taken from the circular of the J. L. Mott Iron Works, of New York. If not objectionable on account of expense, the supply and waste pipes, and the trap, may be of brass, finished or nickel-plated, but a plain, neat job of lead piping will answer very well.

The illustration (Fig. 240) shows a sink, imported by the above firm from England, supported on graceful galvanized or bronzed legs, with a handsome frame on top of the sink, and a marble back.

FIG. 241.—Open Arrangement of Pantry Sink, with Draining Shelf.

The neatest arrangement is to have the floor under the sink—or else the entire kitchen floor—laid with tiles, which may also be carried up along the wall behind the sink. The sink should always be fitted with a high back of iron, glass or marble, to prevent defacing the rear wall by splashing.

A pantry sink may be fitted up in a similar manner, with draining shelf and drawers at one side, but all open directly under the sink, as shown in Fig. 241.

FIG. 242a.—Slop Sink, with Flushing Cistern.

House-maid's sinks should be treated in the same way, but still more important is such a plain arrangement for slop sinks, which otherwise are liable to get very foul and offensive. Fig. 242a shows a slop hopper fitted up with frame and marble back, and a flushing cistern overhead, as sold by the J. L. Mott Iron Works. It would be preferable to have no wood work at all around a slop hopper, everything being in plain sight, open to

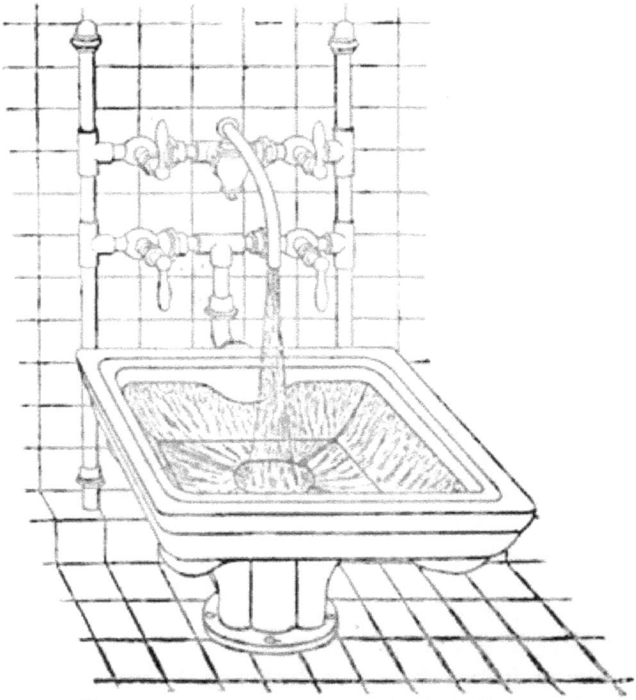

FIG. 243.—Slop Hopper, set on a tiled floor.

inspection, accessible for cleaning and scrubbing. (See Fig. 242b.) Fig. 243 shows such a slop hopper, as sold by The Meyer, Sniffen Co., of New York.

Slop sinks and hoppers should stand in a well-lighted and ventilated closet, or else, where the bath-room is of ample dimensions, in the bath-room. Never should they be placed in a dark closet.

FIG. 242b.—Slop Hopper Sink, without wood-work.

Laundry tubs should also be set on legs and be left open under the tubs, leaving the waste pipe and trap in full sight. Fig. 244 shows the beauti-

FIG. 244.—Laundry Tubs, set on legs.

ful porcelain washtubs, sold by the J. L. Mott Iron Works, set on ornamental legs, with a top frame of hard wood and a back, which may be of marble, through which the faucets for hot and cold water pass. Tiling for the floor of the laundry adds to its beauty and cleanly appearance.

FIG. 245.—Open Arrangement for Wash-basins; slab supported on brackets.

The same principles should be applied to stationary wash basins. In place of the usual cabinet work, let the marble slab be supported on ornamental iron or brass brackets, fastened to the walls (see Fig 245), or else use a pair of marble supports, or a handsome frame, on which the slab rests, supported by bronzed or otherwise decorated iron or wooden turned legs. (See Fig. 246.) If de-

FIG. 246.—Open Arrangement for Wash-basins; slab supported on a frame resting on turned legs.

sired, the trap may be of brass, finished or nickel-plated, and the supply and waste pipes may be similar. Even where householders would object to keeping the pipes and traps in sight, it is possible to arrange a lavatory without the usual tight cabinet-work, as shown in sketch, Fig. 247. The space under the slab and bowl, which latter is of the tip-up type, is left entirely open, and the pipes are concealed behind a movable panel near the rear wall. With a hardwood floor, such lavatory is certainly more cleanly and inviting in appearance than the usual apparatus. Where it is desired to

Fig. 247.—Open Cabinet-work for a Lavatory.

leave all plumbing in sight, and where means are moderate, a handsome job of lead piping, well-shaped wiped joints, etc., are not at all objectionable. Says Mr. Jas. C. Bayles, in describing his ideal house, No. 26 Daydream avenue:

"None of my fixtures are boxed in. I prefer to have everything open and not to make little closets under the fixtures. To my mind there is nothing unsightly about neat pipes with cleanly wiped joints. I like to look at them when everything is as it should be. Besides, I know that these little closets are nothing but poke-holes for old shoes, dirty cloths, musty wooden pails, and other bric-a-brac which properly belong in the ash-barrel. The only way to prevent such accumulations is to have no place where they can accumulate. I let the plumbers who did any work know that nothing was to be covered, and that all the woodwork I should have about the basins and closets was just what was needed to hold up the slabs and seats. They could not understand why I fancied such an arrangement, but finding that I had made up my mind to do as I said, they did their work with extra neatness, and when they had it finished I believe it gave them a positive pleasure to look at it. I forgot to mention that some of my pipes are run inside the walls or partitions. . . . I like to take a look at my pipes occasionally."

FIG. 248.—Hip-bath standing on the floor.

To set bathtubs of all kinds in an open manner, is quite customary in Europe, with the heavy copper tubs usually adopted.

Fig. 248 illustrates a handsome but expensive hip-bath, imported by The Meyer, Sniffen Co., and fitted without any woodwork whatever, all valves and pipes being in plain sight. The American copper-lined tubs require, of course, some exterior finish in woodwork. But enamelled iron bathtubs, and those of earthenware ("Imperial" and "Royal" porcelain tubs) can dispense with woodwork.

Fig. 249 shows Mott's iron enamelled tub set

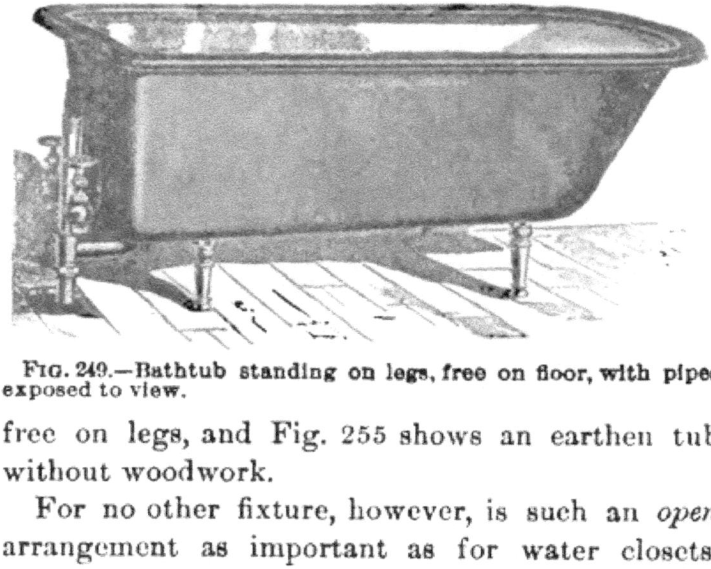

FIG. 249.—Bathtub standing on legs, free on floor, with pipes exposed to view.

free on legs, and Fig. 255 shows an earthen tub without woodwork.

For no other fixture, however, is such an *open* arrangement as important as for water closets. These should have no other woodwork but the seat ; a riser can always be dispensed with. Even closets with machinery, consisting of an iron body and earthen bowl, have nothing objectionable in appearance if fitted up in this manner (see Fig. 250).

The best closets—all earthenware bowls without any marble parts—look most handsome if set on a floor of white tiles, the back and sides of the closet being similarly tiled, and often having a dado of ornamental or colored tiles. In this case, the

FIG. 250.—Arrangement of cabinet work for a closet with movable parts (Hygieia closet).

seat should only be a board of ash, oak or mahogany, well finished and polished, hinged at one end or at the back, and turned up when not in use (see Fig. 251 and Fig. 255). There is no necessity for any further cover, and arranged in this way, hopper or washout water closets may well take the place of slopsinks and urinals.

FIG. 251.—Brighton water closet seat.

In regard to this, we find in a recent volume, "Our Homes, and How to Make Them Healthy," the following advice :

"Another point deserving of consideration by every one about to fix a new water closet apparatus, is the arrangement of the seat and the enclosure of the apparatus. The apparatus is usually fixed and enclosed, so that in course of time a vast amount of dust and dirt accumulates beneath the seat, or, indeed, may have been left there by the workmen when the closet was built;

and where the closet is used for emptying slops of any kind, it commonly happens that small quantities of liquid are allowed to splash on the top of the basin—not sufficient, perhaps, to run away, but to keep a certain amount of permanent dampness on the floor of the space beneath the seat, and to give to the entire closet a constant smell. It would go far to promote cleanliness and prevent this smell if the seat enclosure were wholly dispensed with, and the floor, with its carpet,* or oilcloth, were continued entirely under the seat. In the case of all the best kinds of closet apparatus, comprising merely a basin with siphon trap beneath—all in one piece of glazed stoneware—there would be no eyesore in such an arrangement, while every nook and corner would be visible, and subject to the frequent application of the broom and duster."

FIG. 252.—Earthen Hopper, with wooden rim.

Fig. 252 shows the simplest possible method of fitting up a closet with seat. A well-finished hardwood rim is placed and fastened on top of the hopper, and the latter may be set on a tiled floor

* A carpet should not be recommended.—W. P. G.

or on a slab of best quality slate. This arrangement is especially adapted to work-shops, factories, railroad stations, hospitals, etc.

Fig. 253 illustrates a well-known hopper closet (Rhoad's) made of earthenware, the top being shaped so as to serve as a seat, thereby dispensing entirely with any wood-work, which is always more or less absorbent and becomes in time saturated with urine and perspiration from the body. If this hopper stands in a well-heated apartment, it has much to recommend it, especially for hospitals. If placed in a room not well warmed in winter time, the closet is liable to become filthy through improper use.

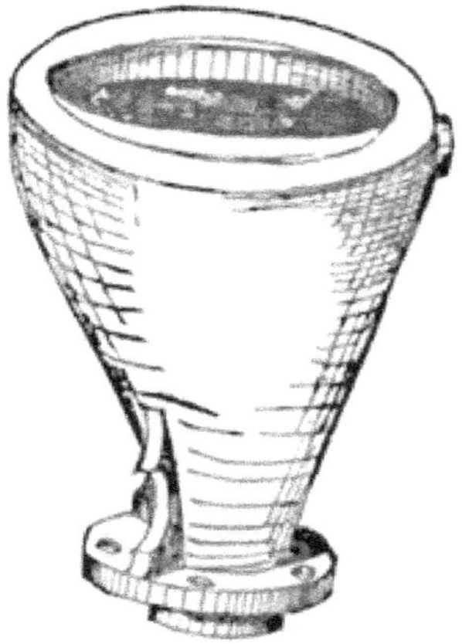

FIG. 253.—Porcelain Seated Hopper.

In contrast with the two closet seats just described, Fig. 254 (taken from the J. L. Mott Iron Works' Catalogue of Improved Water-Closets) illustrates an elegant and ornamental seat, supported on bronzed, galvanized or gilt iron legs. The seat is further fitted with a porcelain drip tray placed just on top of the bowl, which allows the closet to be used as a urinal or slop sink. The hard-wood seat fits closely over the porcelain safe. As the appearance of the hole in the seat is to many still objectionable, there is a handsome cover, but riser and side pieces are dispensed with, exhibiting freely the cleanly tiled floor and walls.

Fig. 255 is a sketch, illustrating the general appearance of a bath-room, arranged according to the principles given. As shown in the illustration, the entire floor is made water-tight, and finished with tiling laid in concrete. The floor directly under all fixtures is somewhat lower than the floor in the centre of the bathroom, and both are joined

Fig. 254.—Mott's Earthen Hopper, with cabinet-work, leaving all parts of closet exposed.

by an easy slope, which is shown in the sketch. Any drippings or spatterings on the main floor will run down the slope to the lower floor. This latter takes the place of the usual safes under fixtures, and has a pitch from all sides to one point (at the left of the bath), at which is arranged a drip pipe, covered with a plated strainer, running vertically down to a sink in the basement, over which it discharges.

The fixtures shown have little or no woodwork about them, the lavatory being supported on artistic brass or bronzed brackets; the closet, a

FIG. 255.—Sketch of a Bath-room, fitted up in an open manner, with tiled floor.

porcelain basin or hopper standing free on the floor, has only a hardwood seat, turned up against the wall, if the closet is not in use; the seat rests, if turned down, on two cleats, supported by bronzed or brass legs. The bath-tub of earthenware stands on short legs, the whole space under the tub being free of access. All supply and waste pipes are in plain sight. The bath-room has a large window, opening to the outer air, and proper provision is made for ventilation in winter time by means of a foul air exit flue. To secure comfort in winter time, the bath-room is heated by indirect radiation (by means of steam or hot water coils in the basement), and a plentiful supply of pure air, moderately heated, introduced through a register in one of the walls (not shown in the drawing).

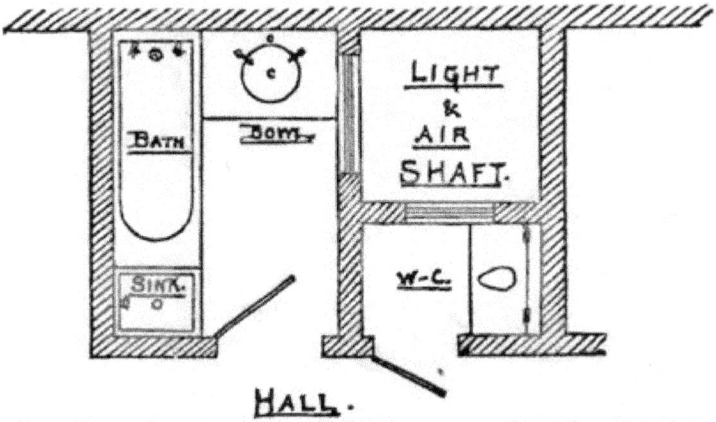

FIG. 256.—Arrangement of Bath-room and Water-closet for City House.

It is only necessary to compare this bath-room with the one shown in Fig. 1, which represents the usual manner of arranging the lavatory, closet and bath-tub in city houses, to understand at once the great advantages of such open arrangement.

Bath-rooms should, wherever possible, be located near an outside wall, with windows affording ample light and ventilation. If they must be located in the center of the house, special ventilation of the apartment must be provided. In regard to this matter we must refer our readers to treatises on "Ventilation."*

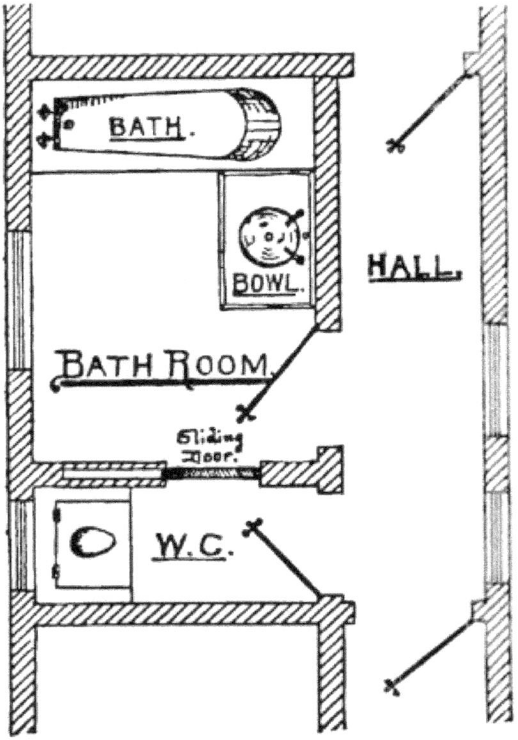

FIG. 257.—Arrangement of Bath-room and Water-Closet for Country Houses.

Speaking of the arrangement of bath-rooms, I wish to state that the American custom of locating the bath-tub, bowl and water-closet in the same apartment, is, in my judgment, objectionable, and

*A very readable account of "House Building in its relation to Hygiene," especially "Heating and Ventilation," is given by Carl Pfeiffer, Esq., Architect, in "Wood's Household Practice of Medicine," Vol. I.

should only be adopted in the case of large residences, having a great number of bath and dressing-rooms. For small dwelling-houses, cottages, and for apartment houses, the water-closet should be located in a separate, well-lighted and well-ventilated apartment, with a door opening into the hallway, if possible, adjoining the door leading to the bath-room. (See Figs. 256 and 257.)

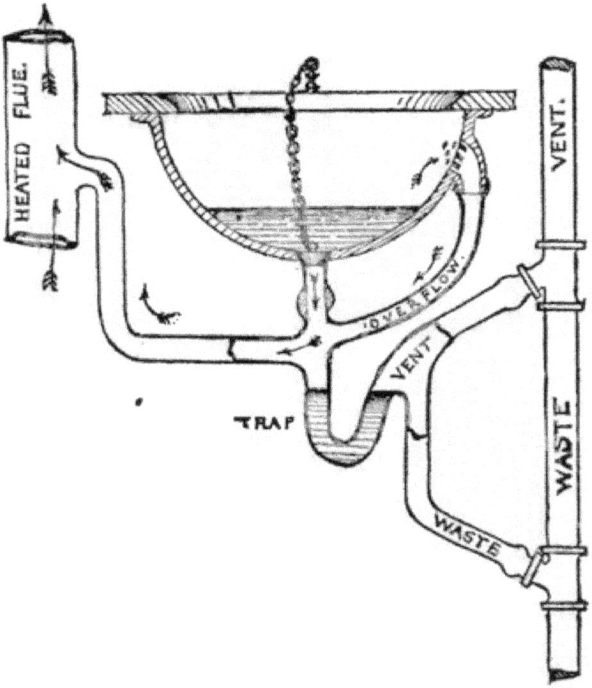

FIG. 258.—Ventilation of House-side of Trap.

A most excellent arrangement for preventing any fouling of the air through plumbing fixtures consists in ventilating not only the soil and waste pipes, and the traps, but, in addition to these, the house-side of the trap or the generally short length of waste pipe between the fixture and the trap, and the overflow pipes, where such are provided. We have already mentioned that overflow pipes, and any waste pipe not often used and flushed, are liable to become foul and ill-smelling, and for overflow pipes in particular such a ventilation is much to be recommended. It consists in running vent pipes of proper size from the house-side of the trap to some *constantly* heated flue or shaft. (See Fig. 258). Such a ventilation will also remove any gases that may possibly be given off from the house-side of a water seal, in case the

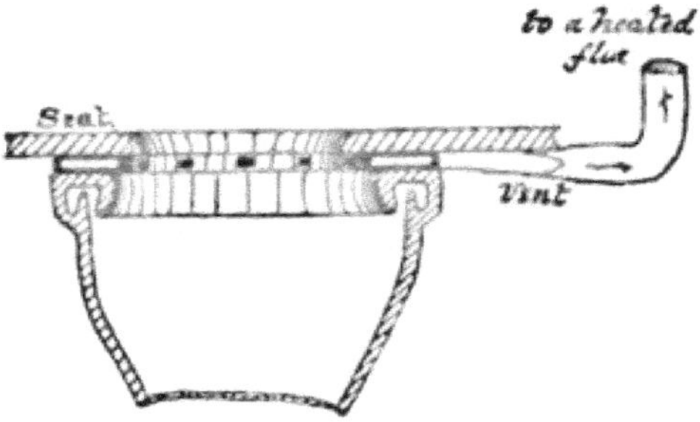

FIG. 258.—Vent of Water-closet under the seat.

water in the latter should become stagnant. In office buildings, stores, factories, it is not a difficult matter to secure such a *constant* draft, by the use of a steam-coil or a smoke-stack, and even in a private dwelling such a ventilation can be provided. We have, heretofore, objected to running either soil, waste or vent pipes into a heated flue, but the arrangement for ventilating fixtures differs from the former, as the ventilation is all on the house-side of the trap. Fig. 245 shows a lavatory ventilated in this manner. Figs. 259, 260 and 261 illustrate this method of ventilation applied to water-closets. The ventilation is arranged either directly under the seat (Fig. 259) or the bowl is provided with a vent pipe attachment (Fig. 260),

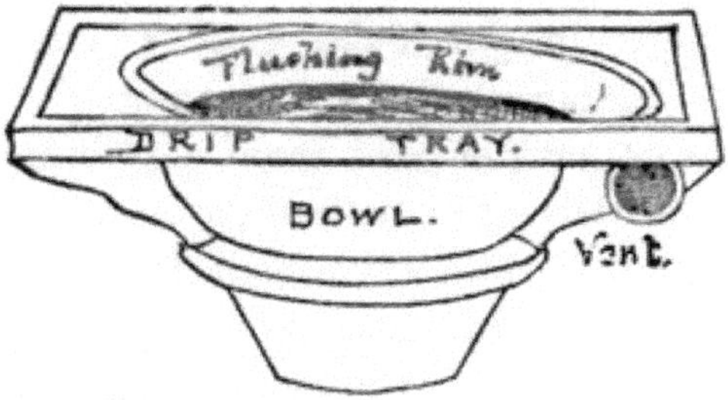

FIG. 260.—Vent of Water-closet from side of bowl.

or else the hopper is provided just above the house-side of the water seal with a vent (Fig. 261).

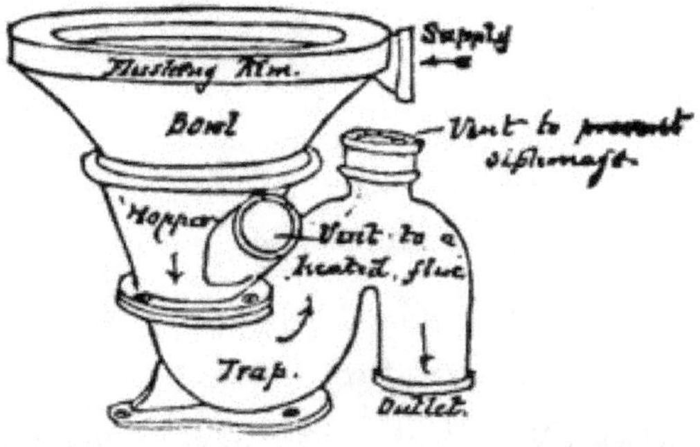

FIG. 261.—Vent of Water-closet from under the hopper.

In Fig. 261 such ventilation by means of a gas-jet is indicated, the pipe B being the vent leading from the closet bowl to the flue.

Such vent pipes from fixtures will secure a constant draft down through strainers and overflow of basins, baths and sinks, as well as water-closets, and will, to some extent, assist in changing or removing the vitiated air of the apartment, instead of, as is usually the case, being the cause of the pollution of the air in the rooms. For bath-rooms, without an external window, this would, *per se*, be insufficient, and a special ventilation of the room should be arranged as already stated.

I do not agree with those that would banish all modern conveniences from the main portion of the house, and would place them all in an annex, cut off from the main house. The comfort and convenience of most plumbing fixtures would, to a great extent, be lost by following such a plan. What should be done is to abolish water-closets and wash-bowls from sleeping-rooms, or unventilated closed closets, adjoining these—as is so often found in American hotels. I believe, on the other hand, that it is quite possible to arrange a bath-room in the centre of a house, or adjoining a sleeping-room, in such a manner as to be perfectly healthy.

Mr. E. C. Gardner, the well-known architectural writer, thus defines a bath-room for a "home":

"A bath-room, with all the plumbing articles it usually contains, must possess at least three special characteristics. It must be easily warmed in cold weather, otherwise the annual bill for repairs will be greater than the cost of coal for the whole house; its walls, floors and ceilings must be impervious to sound. The music of murmuring brooks is delightful to our ears, so is the patter of the soft rain on the roof; but the splashing of water in a bath-tub and the gurgling of unseen water-pipes, are not pleasant accompaniments to a dinner-table conversation. Thirdly, it must be perfectly ventilated—not the drain pipes merely, but the room itself—in summer and in winter. Two of the above conditions can best be secured by arranging to have this important room placed in a detached or semi-detached wing; and here begin the compromises between convenience, cost and safety. It is convenient to have a bath-room attached to every chamber, and there is no doubt that this may be done with entire safety, provided you do not regard the cost. In your plan I have adopted the middle course. There is one bath-room for all the chambers of the second floor, not too remote, but somewhat retired, and having no communication with any other room. It is ventilated by a large open flue carried up directly through the roof; it has also an outside window and inlets for fresh air near the floor. All the walls and partitions around it will be double and filled with mineral wool, and the floors will be deafened. The 'house-side' of the water-closet traps will have three-inch iron pipes running to the ventilating flue beside the kitchen chimney, a flue that will always be warm, and therefore certain to give a strong upward draught at all times, which cannot be said of any other flue in the house, not even of the main drain or soil pipe which passes up through the roof. It would be easy to keep other flues warmed in cold weather by steam pipes, but in summer you will have no steam for heating purposes. A 'circulation-pipe' might be attached to a boiler on the kitchen range for this purpose, but in the present case such a contrivance would cost more than the iron pipe carried from the bath-room to the flue that is warmed by the kitchen fire. A good way to build this ventilating flue is to enclose the smoke-pipe from the range, which may be of iron or glazed earthen pipe, in a larger brick flue or chamber, keeping it in place by bars of iron laid into the masonry. The rising current of warm air around the heated smoke-pipe will be as constant and reliable as the trade winds."

We have already stated, in speaking of fixtures, that overflow pipes are liable to become a nuisance and should be dispensed with wherever possible. We have also endeavored to offer practical suggestions having this object in view.

It has been customary, hitherto, to provide "set fixtures" not only with overflow pipes, but, in addition to these, with "safes" or linings of sheet-lead on the floor, turned up two or three inches, from which a drip pipe carries any leakage of water safely away, thus preventing damage to ceilings. Such drip pipes should, under no circumstances whatever, be connected to any soil or waste pipe, or any sewer. They must run vertically down to the basement or to the cellar, discharging over an open sink, or into a movable pail, or they must stop at the ceiling of cellar, their mouth being closed with a return-bend, having a deep water-seal, to prevent cellar air from rising to the upper floors.

Where such "safes" and drip pipes are in use, we would propose to discharge overflow pipes over such safes; the drip pipe will, then, act as an overflow pipe, carrying any overflowing water to the open sink in the cellar, which in its turn has a trapped connection to the soil pipe or house sewer.

Lead safes, however, are a very unsightly addition to any plumbing fixture. Our suggestions in regard to setting fixtures in an "open" manner could hardly be followed, where lead safes are used. On the other hand, it seems at least very doubtful if such safes are really needed where fixtures are set without woodwork. In the latter case a leak at a coupling of a washbowl faucet, or any other leak, would speedily be detected before doing much damage. We therefore suggest doing away with "safes" as much as possible.

The following design for a bathroom would seem, although more costly, to be superior to the usual arrangement. (See Fig. 255.) Let the part of the floor on which the fixtures stand be made *impervious*, either by making it a cement or concrete floor, or finishing it up with broken stones of various colors and with designs (so-called Terrazzo work), or else by laying white Minton or marble tiles. The latter arrangement is, of course, the most expensive, but nothing can surpass such a finish of a bathroom in point of appearance and cleanliness. The walls may also be finished, to a greater or lesser height, in tiles, decorative tiles being used extensively for such purpose. Let the floor on which the fixtures stand be somewhat lower than that of the remainder of the room. Let the impervious floor be graded toward one outlet (shown in the sketch), and carry from it a drip-pipe to an open sink in the cellar, carefully protecting the outlet in the floor by a plated strainer. Now abolish both overflow pipes and lead safes. Be sure that you have a good workman who understands the practical details of his craft. Let him do the work *right* in the first place, and there will be little, if any, danger from leaks or overflows.

The first outlay for such an arrangement will be greater than if the work were done in the usual manner, but the subsequent repairs, leakages, annoyances in calling in the plumber, will be much less frequent. Yet we venture to say that very few architects would be willing to depart from the old and defective methods of doing things. It is all the more agreeable, therefore, to find an occasional exception. In a book full of sound, practical advice on matters connected with "Building a Home," the well-known architect, Mr. E. C. Gardner, writes as follows:

"It is customary, and doubtless wise, considering our habit of doing things so imperfectly the first time, that we have no confidence in their stability, to place large basins of sheet-lead under all plumbing articles, lest from some cause they should 'spring a leak' and damage the floors or ceilings below them. One strong safeguard being better than two weak ones, I would dispense with the 'overflow' and arrange so that when anything ran over accidentally the lead basin or 'safe' should catch the water and carry it through an ample waste-pipe of its own to some inoffensive outlet. This would, perhaps, involve setting the plumbing articles in the most simple and open fashion—which ought always to be done. *'Cabinets,' cupboards, casings and wood finish, no matter how full of conveniences, or how elegantly made, are worse than useless in connection with plumbing fixtures, which, for all reasons, should stand forth in absolute nakedness. They must be so strongly and simply made, that no concealment will be necessary.*" (The italics are mine).

And now, a few words in regard to the care of plumbing work. No matter how well planned and arranged, plumbing fixtures must be judiciously used, and require looking after from time to time. Even the best ventilated and best flushed water-closet will get dirty and ill-smelling, unless often cleaned; the same is true of kitchen sinks, laundry tubs, slop-hoppers, and other appliances. It is especially important that the water in all traps must be frequently changed. Reference to this point was made on page 58, and further on (page 268) an apparatus will be mentioned having this end in view. A good house-wife will instruct the house-maid in regard to these cleaning operations, which should be repeated at fixed and frequent intervals, certainly as often as once a week.

All earthenware should be thoroughly cleaned by means of hot water, soap, and a scrubbing-brush; dust and dirt must be removed, and the floors and walls frequently washed and scrubbed. All this will be much facilitated, first, by arranging the fixtures iu an open manner, as described above, and second, by locating each fixture in a well-lighted apartment or closet. A valuable addition to a well-arranged bath-room, such as shown in Fig. 255, will be a bibb with screw nozzle, to which a hose can be attached, located near the outlet in the impervious floor, shown in the above cut, by means of which hose a thorough scouring of the sides of the water-closet bowl, of the earthen bath tub, of the tiled floor, etc., can be effected.

In addition to such regular cleaning operations, inspections of the whole plumbing work are much to be recommended. It is true, that if all pipes are kept in sight, inspections are not often required.

A prudent householder will, nevertheless, examine the plumbing work at least as often as the annual house-cleaning occurs, to assure himself of the reliability of all traps under fixtures, of the good condition of all flushing apparatus, of the absence of leaks, etc.

The best disinfectant in all cases is fresh air and a sure and bountiful flush of water, assisted by manual cleansing. It may, however, at times become necessary to use disinfectants for those plumbing fixtures receiving discharges from the human body. A diluted solution of bichloride of mercury has been recommended lately, as being the best; it must be used with great caution, as it is a strong poison. Sulphate of iron or copperas is much cheaper; both should be followed with a large quantity of clean water, to prevent a chemical action in the waste pipes and traps.

Considerable trouble is experienced in the proper care of plumbing in dwellings occupied only during a part of the year. There are first a large number of city residences, which are generally closed, or at least their plumbing fixtures put out of use for two or three months, and sometimes for a longer period during the Summer. The great danger in this case is from evaporation of the water in traps, the seal of which is rarely more than 1½ or 2 inches in depth. To quote from T. M. Clark, Esq., Architect:

"Few people need to be told that a week or two of hot weather is enough to evaporate the sealing water from the traps of wash-bowls, baths, or even water-closets, leaving an open passage from the drains into the house, through which sewer vapors flow freely, often saturating curtains, carpets, and furniture with their faint, sickly odor, to salute the family on its return home in the autumn. When we reflect, also, that the reopening of the house usually takes place in the most fatal month of the year—September—when the system is especially susceptible to zymotic influences, and that the return of delicate persons from the country air to the stifling atmosphere of the city, is generally attended with a certain depression of the vital powers, the danger of sudden exposure to the influence of a house where foul vapors have for months been floating undisturbed, and their deposits accumulating and corrupting in the darkness, is evident, and the trifling care which is necessary to give reasonable security against at least the unchecked circulation of foul air in the rooms will be well repaid."

The ordinary water-seal trap affords no protection in case of evaporation of the water; mechanical traps with a flap-valve, a trap with a gravity ball-valve, or with floating ball, and mercury-sealed traps are preferable in this respect (See Chapter V.)

Fixtures with a socket and plug, or a waste cock (bowls, bath-tubs, pantry sinks, wash-tubs), may be closed against sewer air by shutting the waste-valve, or closing the outlet with the plug, and filling the fixture with water. The holes for the overflow-pipe are generally closed with corks, in the case of bowls, and with paper secured with glue, over the outlet, in the case of sinks and tubs. This does not, of course, afford perfect protection.

The common open strainers for sinks can be replaced by plug strainers, and the latter inserted and the sink filled with water.

For such a case, fixtures without overflow pipe, such as proposed on page 262, offer great advantages.

A much better, though more costly protection, may be found in providing each waste pipe with a lever handle round way stop-cock, to be shut off when the house is being closed. It need hardly be mentioned that the water supply must be shut off

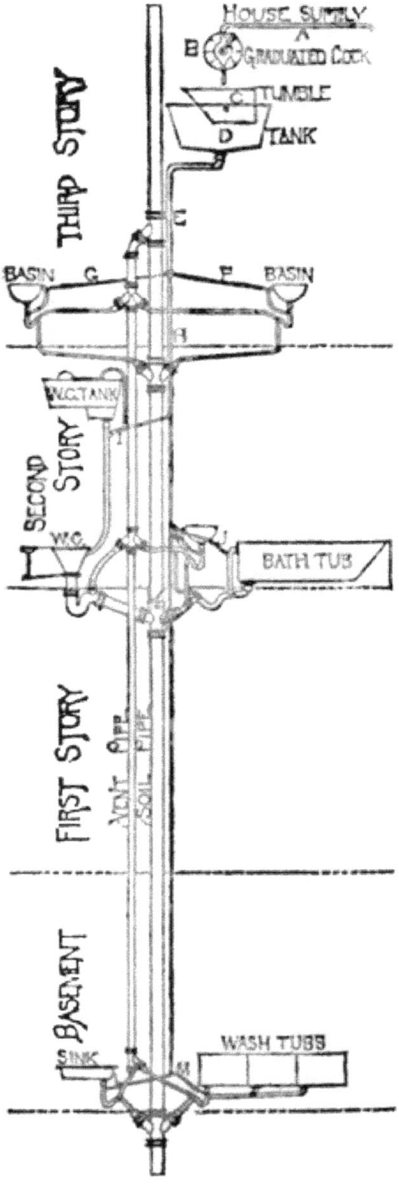

FIG. 262.

Apparatus for maintaining the seal in traps.

from every fixture before closing the stop-cock on the waste pipe.

For the traps of water-closets, the only remedy would be to dip out all water and replace it by oil or by a solution of chloride of calcium. Even a piece of rock-salt, placed into the water of the trap, will tend to keep it filled by absorbing moisture from the air. These latter solutions may also be used for all traps on smaller waste pipes.

Recently only, an apparatus for maintaining the seals of traps came to my notice,* invented by Louis M. Hooper, Esq., C.E. It is illustrated in Fig. 262.

To the house supply is attached a graduated cock B, from which water dribbles into the tumbling tank C, to be discharged at more or less frequent intervals into a vessel D, from which a pipe with graduating branches carries the water to all fixtures in the dwelling, those having overflow-pipes receiving their water through these, while water-closets receive their share through a branch entering the supply pipe between the cistern and the bowl, and kitchen sinks and wash-tubs receiving their amount into the waste pipe on the house-side of the trap. Such an arrangement would render the whole trap system safe against evaporation, and, although adding to the cost of plumbing, seems well worth considering. It offers the further advantage that it refills the traps under fixtures in case of siphonage (although vent pipes are also added to prevent dead ends in the waste pipe system), and finally, it affords means of

* See *Sanitary Engineer* of Nov. 15, 1883.

automatically changing the contents of the traps under sinks, tubs and bowls, which have no special flushing cistern, and thus will largely contribute to the purity of the air of rooms, containing such plumbing fixtures. (See remarks, pp. 58 and 264.)

Still more trouble is experienced in the case of country or sea-shore dwellings, summer hotels, etc., which are closed during the winter season. The chief danger to the plumbing work arises in this case from the freezing of the water in pipes and traps.

As to supply pipes in such buildings, these should, of course, be planned and laid out in such a manner that every line of pipe can be completely drained and emptied.

All waste pipes, on the other hand, have, in a good system of plumbing, sufficient fall to insure the running off of all water from the pipes.

The difficulty arises from the water which forms the seal in the water-seal traps. In a well-arranged system every trap is, if not in plain sight, at least easily accessible, and every trap may be emptied either by removing a brass trap screw at its bottom, or else by dipping the water out with a sponge. After this is done, each fixture affords an opening to the entrance of sewer gas, and must be closed in the same way as stated above for dwellings closed in summer time.

It is preferable not to empty the traps, but to throw a large quantity of rock-salt into them, which, though it does not render freezing impossible, still renders it much more difficult. If not too expensive, a mixture of glycerine and water may prove of great service.

CHAPTER XII.

REMOVAL AND DISPOSAL OF HOUSEHOLD WASTES.

External Sewerage of Dwellings.

FROM a point about ten feet outside of the cellar walls the house sewer need not be of iron, but may consist of strong, vitrified earthen pipe, or of cement pipe, unless the sewer passes near a well or spring, in which case iron pipe is preferable.

For ordinary-sized dwellings and lots, a pipe sewer four inches in diameter will prove ample to remove the house sewage and the largest rain-fall. Remember that the more the size of the drain is restricted within the limits of desired capacity, the more *self-cleansing* will it be. I take the following useful table of sizes of drains from Denton's "Handbook of House Sanitation:" *

---

*Another reliable table for calculating the size of house drains, is given in W. P. Gerhard's "House Drainage and Sanitary Plumbing," published by D. Van Nostrand, N. Y. 2d Edition. 1884.

See also the "Diagram of Sewer Calculations," constructed by the author and published in 1882, by E. & F. N. Spon, London and New York.

Diameter of Pipe in Inches.	Velocity 3 ft. per second.		Velocity 4½ ft. per second.		Velocity 6 ft. per second.		Velocity 9 ft. per second.	
	Fall.	Discharge in U. S. gals. per minute.	Fall.	Discharge in U. S. gals. per minute.	Fall.	Discharge in U. S. gals. per minute.	Fall.	Discharge in U. S. gals. per minute.
3	1 in 69	64.8	1 in 30.4	97.2	1 in 17.2	129.6	1 in 7.6	194.4
4	1 in 92	115.2	1 in 40.8	172.8	1 in 23.0	230.4	1 in 10.2	345.6
6	1 in 138	259.2	1 in 61.2	388.8	1 in 34.5	518.4	1 in 15.3	777.6
9	1 in 207	594.0	1 in 92.0	891.0	1 in 51.7	1188.0	1 in 23.0	1782.0
12	1 in 276	1051.2	1 in 122.4	1577.0	1 in 69.0	2102.4	1 in 30.6	3153.6

The discharge given refers to pipes running full. For pipes running only half full, the velocity remains the same, and the discharge is just one-half of that given in the table.

The inclination of the house sewer should be, wherever possible, not less than $\frac{1}{4}$ inch per foot, but even a fall of $\frac{1}{8}$ inch to the foot will cause a sufficient velocity in the drain to remove silt and water-closet matter. If the locality does not afford a chance for such an inclination, a proper flushing apparatus for the house sewer must be provided.

To bring the sewer out of reach of frost, it should be laid at least three feet deep. It must be laid in perfectly straight lines. Wherever changes of direction occur, these should be effected by easy curves, made of bent pipes. Branch drains should enter the main house sewer by Y-branches so as to join the flow of both pipes without causing eddies. Should the house sewer be very long, it is well to provide means for occasional inspections, access pipes or lampholes, at distances of about 100 feet, and manholes at distances of 300 to 400 feet. These will, at the same time, if provided with open gratings, perform the important task of ventilating the house sewer throughout its entire length.

Vitrified pipes are manufactured of some kinds of clay, ground in a mill and homogeneously mixed. The mixture is brought to a press and passed through dies, from whence the pipes issue. Smaller sizes are made in horizontal presses, while the larger sizes should preferably be made in upright presses. The pipes are now ready for the glazing, and here two processes may be distinguished, the salt-glazing and the slip-glazing. In the former process the pipes are subjected to a very high temperature in a kiln, into which some salt is thrown, which creates a flux on the pipe surface. To this latter is largely due the glossy appearance of the pipe; it also renders the pipe more impervious and not so easily affected by acids, alkalines, or sewage gases.

Slip-glazed pipes, on the other hand, are made by dipping the pipes into a peculiar glaze called slip, and then drying them in a kiln.

Good vitrified pipes must be circular, and true in section, of a uniform thickness, perfectly straight (this is very important to insure a good line of pipe), free from any cracks or other defects; they should be hard, tough, not porous, and of a highly smooth surface. The thickness of good earthen pipe should average as follows:

DIAMETER OF PIPE IN INCHES.	3	4	5	6	8	10	12	15	18
THICKNESS OF PIPE IN INCHES.	$\frac{3}{8}$	$\frac{1}{2}$	$\frac{5}{8}$	$\frac{11}{16}$	$\frac{3}{4}$	$\frac{7}{8}$	1	$1\frac{1}{4}$	$1\frac{1}{2}$

Vitrified pipes are made in lengths of two or three feet, either plain (Fig. 263, *b*), or else they are made with a socket end. (Fig. 263, *a*). Many engineers prefer the plain or ring pipe, which is laid with sleeves, as this allows of an easy exchange of a single length of pipe from a pipe line already laid, while, with the socket pipe, it becomes necessary to disturb several lengths. To overcome this difficulty, pipes are also manufactured with half-sockets (Fig. 264a), or else they are made plain at both ends, and are bedded with cement in earthen chairs (Fig. 264b).

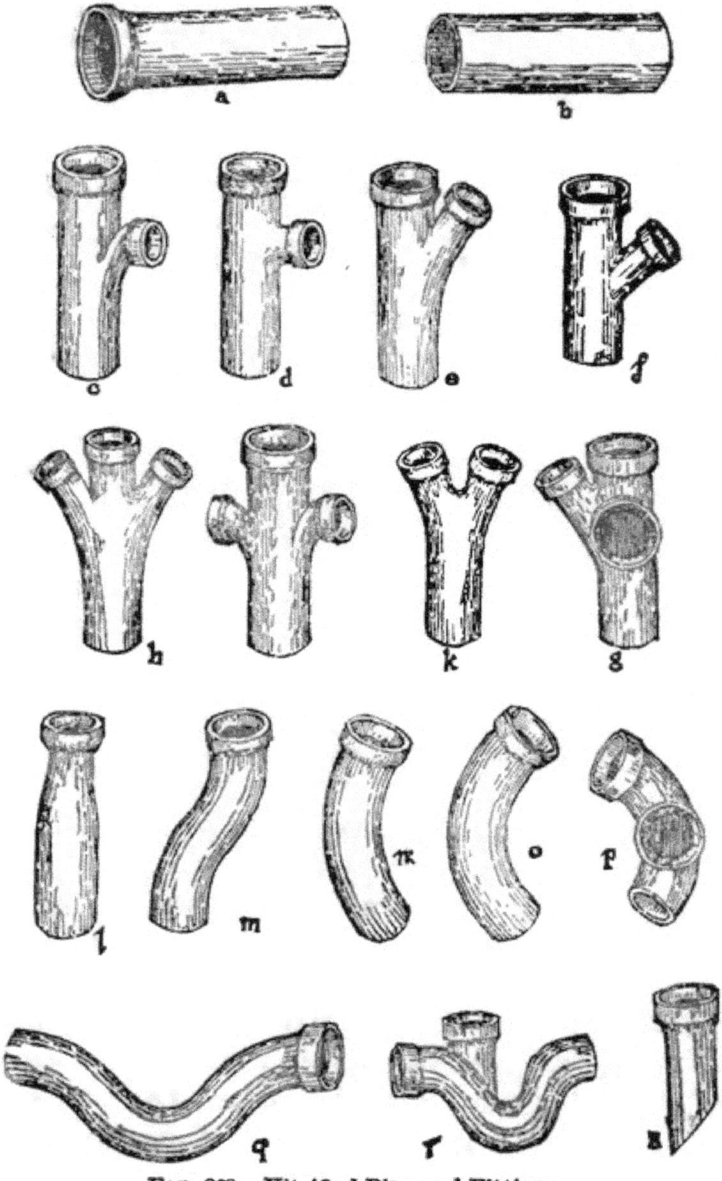

FIG. 263.—Vitrified Pipe and Fittings.

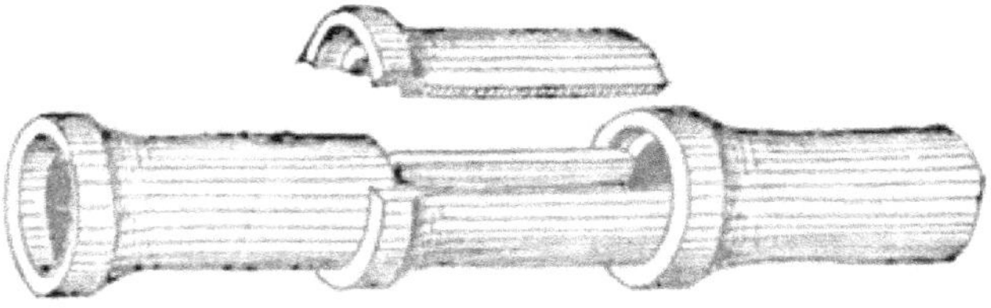

FIG. 264a.—Opercular, or half-socket pipe.

All pipe works manufacture a large number of fittings for earthen pipes, such as traps, Y-branches, T-branches, junction pieces, bends, offsets, etc. (See Fig. 263,*c* to 263,*s*.)

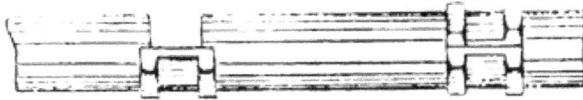

FIG. 264b.—Saddle Chair or Access Pipe.

Cement pipes, though not as universally used as vitrified pipe, have been manufactured for years for drainage purposes. If care is observed in their manufacture from best Portland cement, such pipes can be made very strong and durable, and of a very uniform cross-section. They have also the advantage of not warping, as the earthen pipes do in the kiln. The interior, however, is not as smooth, and unless well flushed they are more apt to become covered with a dangerous slime, dangerous because it will putrefy and thereby fill the pipes with gases of decomposition.

If vitrified pipes are used for sewerage purposes, the pipes must be continuously supported to prevent breakage, and grooves should be cut so as to make the pipe rest on its full length. (See Fig. 265.)

There are but a few so-called "drain layers" who thoroughly understand the laying of pipe sewers. To insure tightness of joints it is well, in using socket pipes, to ram first a small gasket of

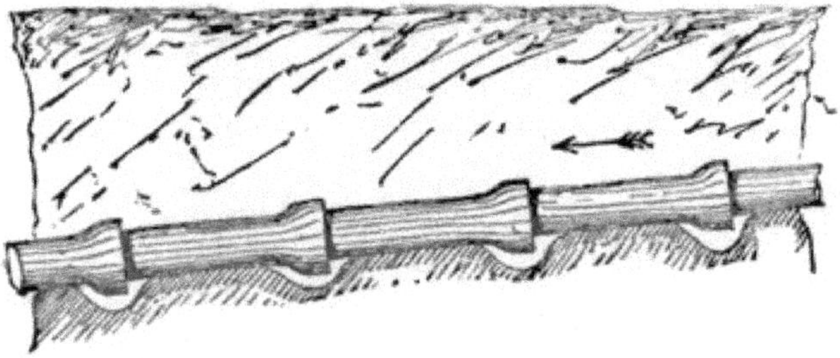

FIG. 265.—Proper method of laying earthen drains.

oakum between spigot and hub, which will prevent the cement from entering at the joints to create in hardening an obstruction. To do so it is quite necessary that the socket should be very deep. The remainder of the space should be filled with a mortar consisting of an even mixture of best

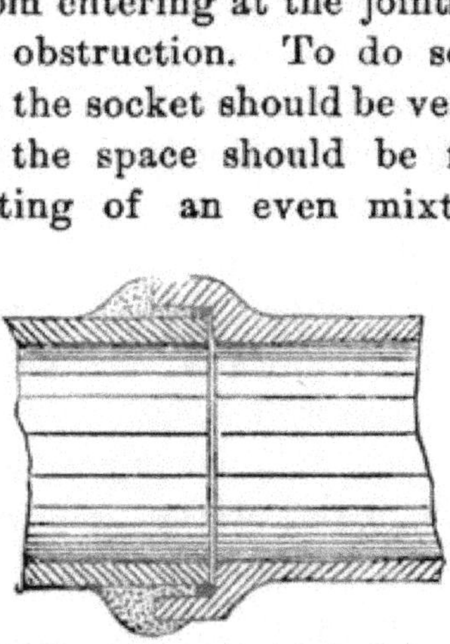

FIG. 266a.—Cement Joint in Vitrified Socket Pipe.

Portland cement and clean sharp sand. The cement and sand should be thoroughly mixed dry, and then wetted up only as needed. No lime should ever be used in this mixture, nor should any cement be used that has begun to set. Cement is also wiped in front of the joint, as shown in Fig. 266a. Fig. 266b illustrates a joint in plain pipe made with a collar or sleeve.

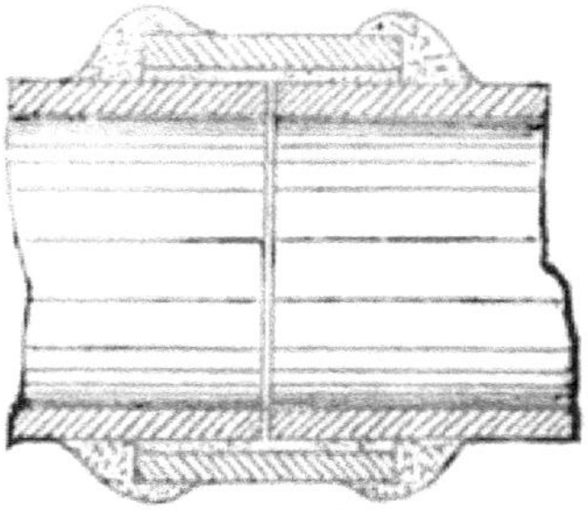

FIG. 266b.—Cement Joint in Vitrified Plain Pipe.

Before refilling the trench it is to be recommended to test the pipes and joints by hydraulic pressure, by closing the main outlet of the house sewer and filling the pipes with water. Considering the usual wretched manner of laying house drains, such tests seem to be extremely necessary. Vitrified pipes with well cemented joints are perfectly able to stand some internal pressure. I recently learned of a pipe-line in Wurtemberg, Germany, 4,020 metres (2½ miles) long, and 10 centimetres (4 inches) in diameter, supplying a railroad tank with 60 cubic metres of water daily, which line was subject, in several places, to a head of water of 8 metres (26.25 feet), equivalent to 11.4 lbs. pressure per square inch. In laying this line of vitrified pipe, each pipe was carefully inspected and tested under 75 lbs. pressure before use. After laying the pipe and after the cement in the joints had hardened, the line was tested in sections, each section being subjected for 15 minutes to a pressure of 60 lbs.

Such severe tests of the external sewerage corresponding to the testing of the internal pipe system by water, will secure work of a quality and character such as is desired for sanitary reasons, namely, a perfectly water-tight conduit, without any joints through which sewage may leak out or sub-soil water enter.

To secure a water-tight joint under specially difficult conditions, such as water in sewer trenches, tides, etc., various pipes have been made with patent joints; for instance, the "Stanford Patent Joint Pipe," which has rings of some bituminous compound cast on the spigot end, and in the socket of each pipe. Just before using them the parts to be jointed are greased, and then the spigot end carefully and truly entered into the socket. (See Fig. 267).

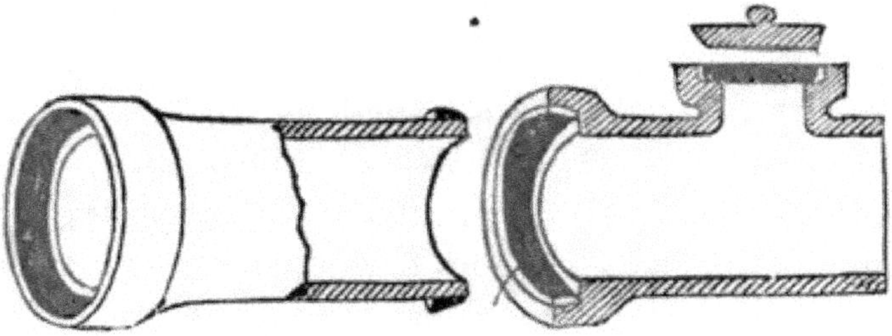

FIG. 267.—Stanford Patent Joint Pipe.

To facilitate future inspections and to remove occasional obstructions, it is to be recommended to keep a correct and detailed record of all drains, their sizes, depths and rate of fall, the location of all traps, Y-branches, man-holes, lamp-holes, vent-openings, junctions, bends, etc.

Disposal of Household Wastes.

By means of the house sewer we effect an instant *removal* of all liquid and semi-liquid wastes from the dwelling. The next, and in some sense most important question, is how to *dispose* of these foul wastes?

In the case of city dwellings we generally find provision made by sewers in the principal thoroughfares. Nevertheless it is only in few cities that sewers have been built as they should be, according to a regular "system", designed and laid out by an engineer of large experience in this special branch of the profession. Much remains to be done in this direction, but the subject of "*city sewerage*" does not properly belong to our volume and cannot be discussed here. Faulty connections between street sewers and house sewers have been mentioned on page 155. Such junctions should be made by competent workmen only, according to rules and under the supervision of an inspector, employed by the city. Wherever a new system of sewers is being built, it is a good practice to provide special house connection pieces for every lot and dwelling on both sides along the line of the main sewer. Sometimes the branches for each house are at once run up to the curb line, thus doing away with the usual annoying and detrimental breaking up of the street pavement.

In all such cases where sewers are built by the city, the final disposal of the sewage is a matter in which the city authorities are more directly concerned than the individual householder. The latter's work stops at the junction between house and street sewer.

The case is different in cities, towns, villages or hamlets without sewers. The serious question of how to dispose of the liquid wastes of the household without creating a nuisance presents itself in such cases to every house-owner or tenant.

In towns or villages with houses built closely together, there is scarcely a remedy for the evil, other than to abolish the disgusting and health-endangering cesspools in the rear of the houses, and to build, by united action of the citizens, a complete and well-planned system of sewers.

In suburban or rural districts, and in the case of isolated buildings with ample and suitable grounds about them, the question can, fortunately, be easily solved in most cases, without incurring the expense of building sewers, the proportionate cost of which for each house would be unusually large in the case of scattered dwellings.

A leaching cesspool in a free and porous soil could, often, be used without immediate danger to the house or the occupants for whom it is intended, or to its surroundings, provided it could be located *very far* from it and on a much lower level. But such an arrangement is nevertheless attended with the risk that the liquid sewage, seeping into the subsoil, may reach some subterranean fissure or stratum, along which it would move, to empty finally into a spring or well, often miles away. Outbreaks of typhoid fever, caused by drinking water contaminated in this manner, have often been traced to such a leaching cesspool. The la -

ter should, therefore, be considered as *always* objectionable in the interest of the public health.

The only proper and rational method of sewage disposal in such case is to return to the soil as fertilizers the wastes from the household. This can be done in a variety of ways, but whatever method may be adopted, it should be borne in mind that *the sewage must be applied to the soil before putrefaction begins, that it should be applied on or near the surface of the soil, within easy reach of the oxidizing influence of the atmosphere, and that it should not be applied in such quantities as to saturate the soil; in other words, the sewage must not be too much diluted, and the application of the sewage to the land must be intermittent.*

Both *surface irrigation* and *sub-surface irrigation* have been successfully employed in disposing of household wastes. Fig. 268 illustrates the disposal by surface irrigation adopted by

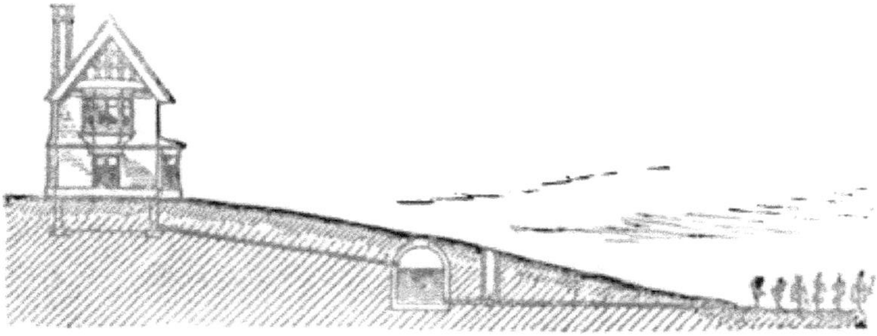

Fig. 268.—Disposal of Household Wastes by Surface Irrigation

Mr. Edward S. Philbrick, C.E., at his country seat at Newport, R. I.* For cottages having little ground about them, this method of disposal

* The illustration is taken from Mr. Philbrick's book, "American Sanitary Engineering."

may become offensive to sight and smell during the hot summer months. The disposal by sub-surface irrigation (see Figs. 269 and 270) is free from these objections, but requires, on the other hand, more work in planning and laying out the system, and it also requires the laying of a net-work of distributing drains laid under and near the surface, which drains occasionally clog up and require taking up, washing out and relaying.

Surface irrigation may be adopted in connection with a small, well-ventilated and perfectly *tight cesspool or sewage tank*, on the top of which is set a small pump, with hose attached, by means of which the liquid may be sprinkled over the lawn or in the kitchen garden. If preferred, a stop-gate may be placed beyond the cesspool, and a drain run from the cesspool to an irrigation field on a lower level, if such can be had. As often as the cesspool is filled, the contents should be discharged by opening the stop-valve. The cesspool should not hold more than a few days' waste water, and should, preferably, contain an intercepting chamber for grease and solids.

As we believe the second method of sewage disposal to be the one preferable for isolated dwellings in most cases, a few words regarding the details of the system may not seem out of place.

This system was first brought into use in England by the Rev. Henry Moule, Vicar of Fordington, the well-known inventor of the earth closet. Sewage disposal by sub-surface irrigation has since been extensively practiced by Mr. Rogers Field and Mr. J. Bayley Denton, both prominent sani-

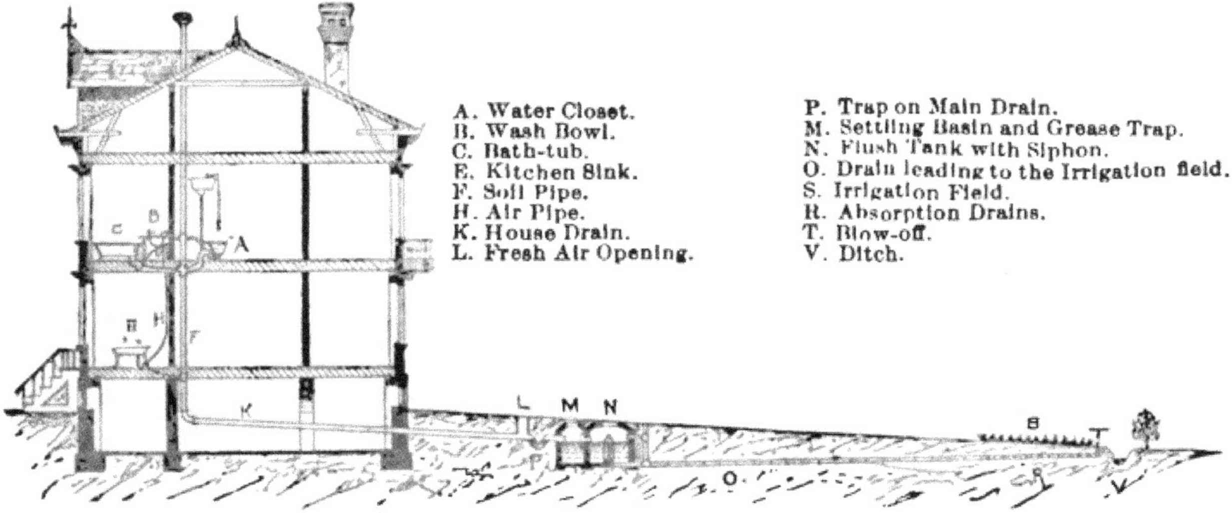
A. Water Closet.
B. Wash Bowl.
C. Bath-tub.
E. Kitchen Sink.
F. Soil Pipe.
H. Air Pipe.
K. House Drain.
L. Fresh Air Opening.
P. Trap on Main Drain.
M. Settling Basin and Grease Trap.
N. Flush Tank with Siphon.
O. Drain leading to the Irrigation field.
S. Irrigation Field.
R. Absorption Drains.
T. Blow-off.
V. Ditch.
A
B
C
E
H
F
K
L
M
N
O
S
T
V

FIG. 260.—Sewage Disposal by Sub-surface Irrigation.

tary engineers in England. To Col. Geo. E. Waring, Jr., of Newport, R. I., is due the credit of having introduced this system in the United States, about twelve ago; first, for his own house in Newport; subsequently for a large number of country houses in the Eastern States; finally, on a large scale, for the disposal of the sewage of the woman's prison, at Sherburn, Mass., and at the Keystone hotel, Bryn Mawr, Pa. It has been adopted since by many civil engineers and architects for the drainage of suburban and country homes, and has received the endorsement of physicians, sanitarians, and Boards of Health.

The principle of the sub-surface irrigation system is briefly this: The porous soil next to the surface has the power of destroying organic substances and rendering them innocuous, partly with the aid of the oxygen contained in the pores of the subsurface, partly by means of the vegetation, since the rootlets of grass and shrubs take up nourishment from these organic matters. The sewage water, from which all impurities have thus been removed, settles away, and becomes still more clarified by filtration, in most cases to such a degree that, if removed by under-drains (land drains), it is found to be quite clear, colorless, free of taste or smell.

All impurities are oxidized and destroyed during the interval between consecutive discharges. The importance of an *intermittent* action becomes, therefore, at once apparent. If this is secured, the upper layers of earth are enabled to take up at each interval between discharges, oxygen from the atmosphere and prepare for the next discharge. Another reason for making the discharge intermittent is to prevent the ground from becoming saturated, wet and swampy.

The cardinal difference between a sub-surface irrigation system and a leaching cesspool is this: In the latter case the amount of soil used for the purification of the sewage is quite small as compared with the former, where the surface can be chosen according to the amount of sewage to be disposed of. The leaching cesspool, too, when newly built, effects some purification and filtration of the household wastes. Soon, however, the pores of the soil clog up, as the organic matter is not completely oxidized at greater depth and as the aid of the vegetation is lost. The soil gradually becomes saturated with sewage matter, which undergoes a slow process of decomposition, during which many unwholesome gases are generated. These gases are given off at the surface and are sucked up into our dwellings, especially in winter time. The other not less serious evil is caused by the sewage soaking unpurified into the ground, thereby threatening to pollute our water supplies.

The sub-surface irrigation system consists essentially of two parts:

First, a tight receptacle for liquid and semi-liquid house refuse, from which the water is discharged at intervals into a system of underground tiles.

Second, a system of common two-inch drain-tiles, laid with open joints, a few inches below the surface of the ground, permitting the liquid sewage to escape at each joint, to be partly purified by the action of roots of grass or shrubbery, partly oxidized by the oxygen attaching to the particles of the soil near the surface.

The construction of the tank depends upon local conditions, such as size of the house, number of inhabitants, character of the foul wastes (slop water only or slop water plus excreta), amount of water used, etc. Several examples will be given later. The size should be regulated so as to have, if possible, one daily discharge, for otherwise the sewage in the tank might commence to decompose, making the tank practically a cesspool. Of course, the tank should, in any case, be located as far away as possible from the dwelling, but the best place for it will depend largely upon the contour of the surface.

The sub-irrigation field should also be remote from the house, if possible in a direction from which the wind would but seldom blow. It should not be located near a well or a spring. It may, in the case of small cottages, consist only of one line of tiles, or it may contain a large number of these, this depending also upon the character of the soil. The system works best in a sandy or gravelly loam, but even in heavy clay soil it has been used with tolerable success. If the land is apt to be wet it must be thoroughly underdrained by a system of deep land drains, otherwise the sewage will soon come out at the surface and convert this into a swamp. Doubt has often been expressed as to the working of the system in winter time. Experi-

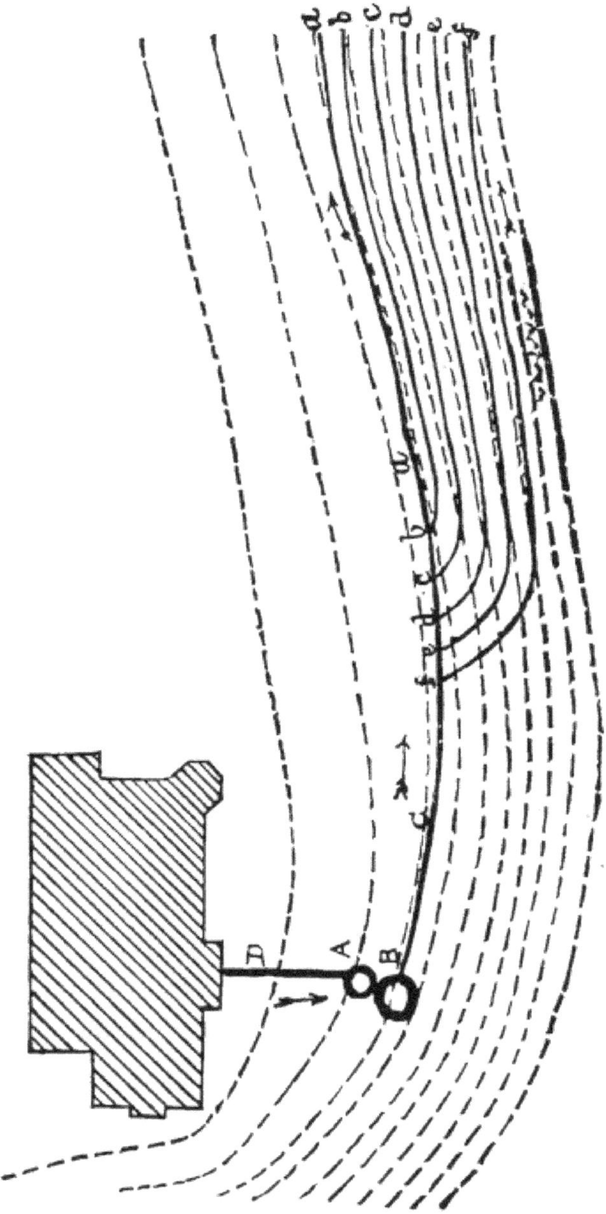

FIG. 270.—Plan of Sewage Disposal by Sub-surface Irrigation.

ence has taught that the distributing tiles laid close to the surface will not freeze, as might be expected, the temperature of the sewage being sufficiently high to keep the sewage in the pipes in motion.

Fig. 270 shows a plan of a sub-surface irrigation system, D being the house sewer, A the intercepting tank, B the flush tank, C the main drain to the sub-irrigation field, and *a*, *b*, *c*, *d*, *e*, *f* the lines of absorption-drains. The dotted lines indicate the contours of the land, and it will be seen that the drains closely follow these lines, thus becoming almost parallel to each other.

The tiles used are shown in Fig. 271, and are

Fig. 271.—Absorption Tiles, with gutters and caps.

common two-inch porous land-tiles, one foot long. They are laid about 8 or 10 inches below the surface on continuous boards, or better in gutters of earthenware, as shown, which gutters must be accurately laid in the trenches at the required grade. If the tiles should clog up, they can be taken up and cleaned, and the relaying into the gutters is then a rather easy matter, which can be accomplished by almost any common laborer. There must be a space left at each joint of about ¼ inch, in order to facilitate the escape of the sewage. To protect the joint from earth or dirt falling from above, small caps are placed at each joint, as shown, over the tiles.

The main line may be 4 inches in diameter, and from it the 2-inch lines branch out by means of Tees or Y-branches, as the case may be, with side openings branching out from the bottom, as shown in Fig. 272. The main line is cemented tightly,

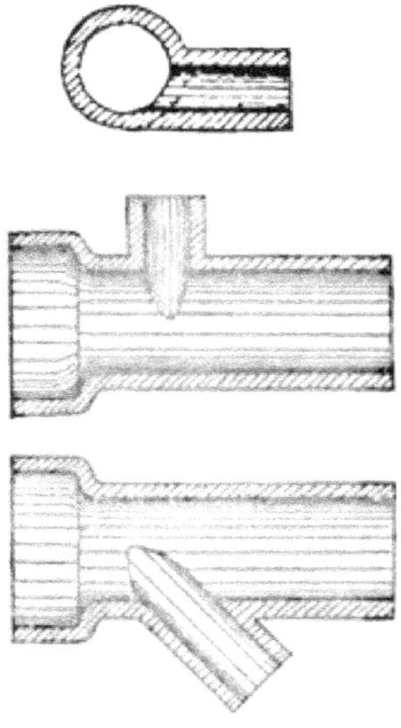

FIG. 272.—Y-branch and Tee-branch for absorption drains.

and so is each branch in the curved part, until it strikes the depth of 8 or 10 inches from the surface (the main 4-inch drain being laid two feet deep or more). The manner of laying the absorption tiles is further illustrated in Fig. 273, and in Fig. 274, showing a cross section through tiles and trench.

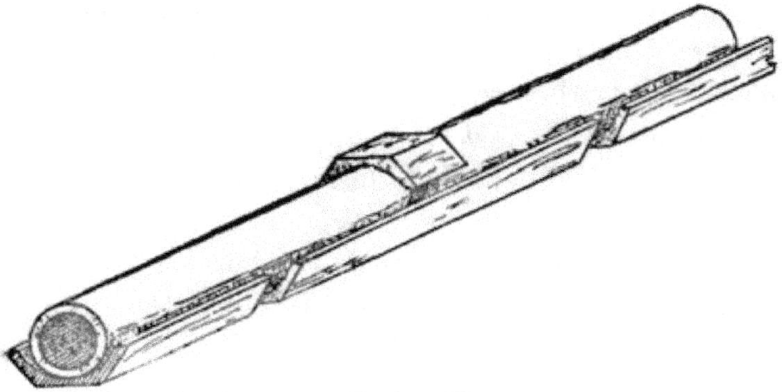

FIG. 273.—Manner of laying absorption drains.

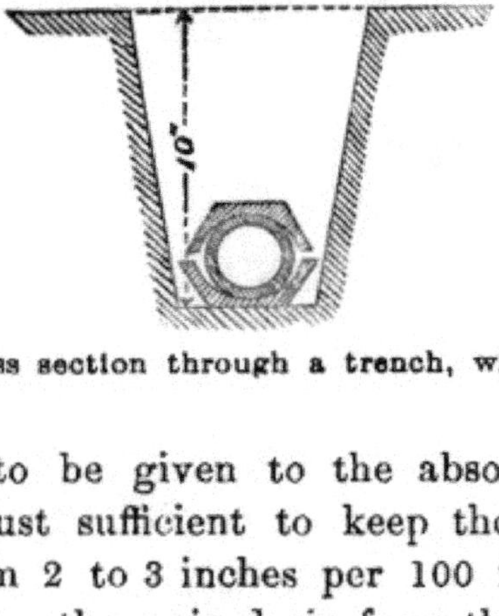

FIG. 274.—Cross section through a trench, with absorption drains.

The fall to be given to the absorption tiles should be just sufficient to keep the sewage in motion ; from 2 to 3 inches per 100 feet is considered ample ; the main drain from the flush tank to the irrigation field may have as much fall as circumstances will permit, but near the absorption drain branches the fall should be limited to 4 or 6 inches per 100 feet, otherwise the sewage would tend to run to the lower part of the field, overcharging the lower lines of drains, and oozing out at the surface. The main 4-inch drain, as well as the 2-inch absorption drains, must be laid with a *perfectly uniform descent*, and much of the success of the system will depend upon the accuracy with which this part of the work is laid out and constructed.

In the case of very small houses, the sewage—consisting of slop-water only—may be distributed by carrying it from the house by hand and pouring it out of a pail into an open hopper or receiver of wood or earthenware with a strainer, from the bottom of which hopper a line of pipe leads to the absorption drains. (See Fig. 275).

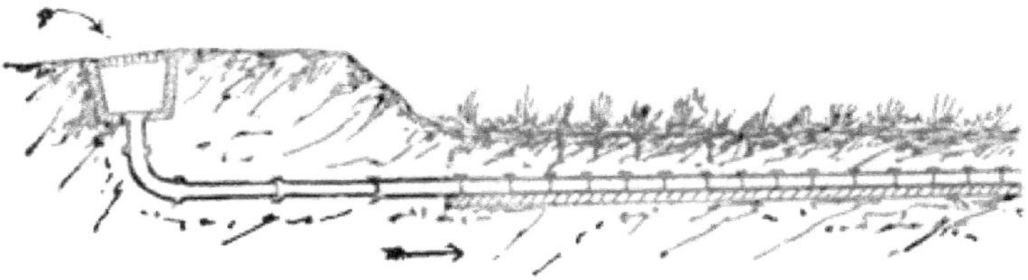

FIG. 275.—Plain hopper for slop-water disposal.

For small cottages having only a kitchen sink, a receiving tank may be built of wood and located at a depth beyond the reach of frost, as shown in Fig. 276, to which runs a waste pipe from the sink. If filled, the tank may be emptied by hand, and thus an intermittent discharge established. The illustration shows a ball float, which is merely intended to open the outlet automatically in case of forgetfulness of the occupant of the cottage. That part of the tank, which is divided from the main tank by a partition, serves as a grease-trap to prevent grease from the kitchen sinks clogging the absorption tiles. In both cases illustrated it is

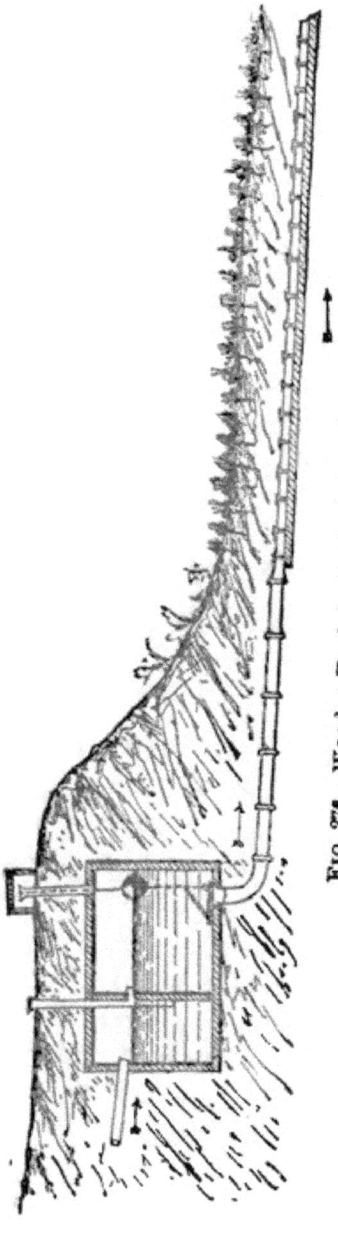

FIG. 276.—Wooden Tank for Slop Water Disposal.

supposed that no water-closet exists in the house. In place of the objectionable privy, there should be some kind of earth-closet, the contents of which should be frequently removed and dug into the ground.

The limits of this book do not permit us to discuss in detail all dry-methods of excrement removal. We have condemned the usual filth-reeking privy as entirely unfit and highly dangerous to health, and offer in the *earth-closet* a substitute, simple and cleanly in operation, entirely inoffensive in use, and well adapted to prevent the unhealthiness of cottage occupants caused so frequently by emanations from accumulations of putrefying excreta. As we limit our remarks to the disposal of excrements of single cottages only, we cannot discuss the extent of the applicability of the dry-earth system to villages or large communities.

"The Dry-Earth System," to quote from Dr. Buchanan's official report made in 1869, "consists in the application, with the greatest procurable detail, of dry earth to fresh human excrement, and in the subsequent removal and use of the mixture for agricultural purposes." The use of dry earth for disposal of excreta, although known in a general way since centuries, originated practically with the Rev. Henry Moule, Vicar of Fordington, the same who devised the sub-surface irrigation system for the disposal of slop-water. *Dry earth* possesses, in a high degree, the power of deodorizing and disinfecting human excreta. A pound and a half of dried and finely-sifted earth is considered sufficient for the average dejection. The quality of the earth used is of great importance. Gravel and sand are useless in this respect, chalk is not adapted to this purpose, while clay is quite a fit material. But the best earth is that of a loamy character, such as garden earth or vegetable humus, which already contain some organic matter. The same quantity of earth can be used over again several times, provided it is thoroughly dried.

Numerous mechanical arrangements have been devised to throw earth in proper quantity and in the right manner upon the excreta deposited in a reservoir under the closet seat. We believe the simplest arrangement for the use of small cottages to be the one shown in the sketch, Fig. 277, of having in the closet a box containing dried and well-sifted earth, which is thrown upon the excreta by means of a hand-scoop after each use.

The excreta should fall into a plain box or pail, or else into a tank on wheels under the seat. The sketch, however, shows a tightly-cemented vault, entirely above ground, open and accessible for cleaning out at the rear, from where the fertilizing mixture should be removed at frequent intervals to be dug under the ground.

It is decidedly preferable not to locate an earth-closet inside of a dwelling. Unless very strict attention is paid to the apparatus, it is apt to become offensive to the smell. A plain shed may be erected, quite close to the rear of the house, if desired, and accessible by means of a covered walk, to prevent exposure in cold weather. Particular care should be taken that no rain-water drips into

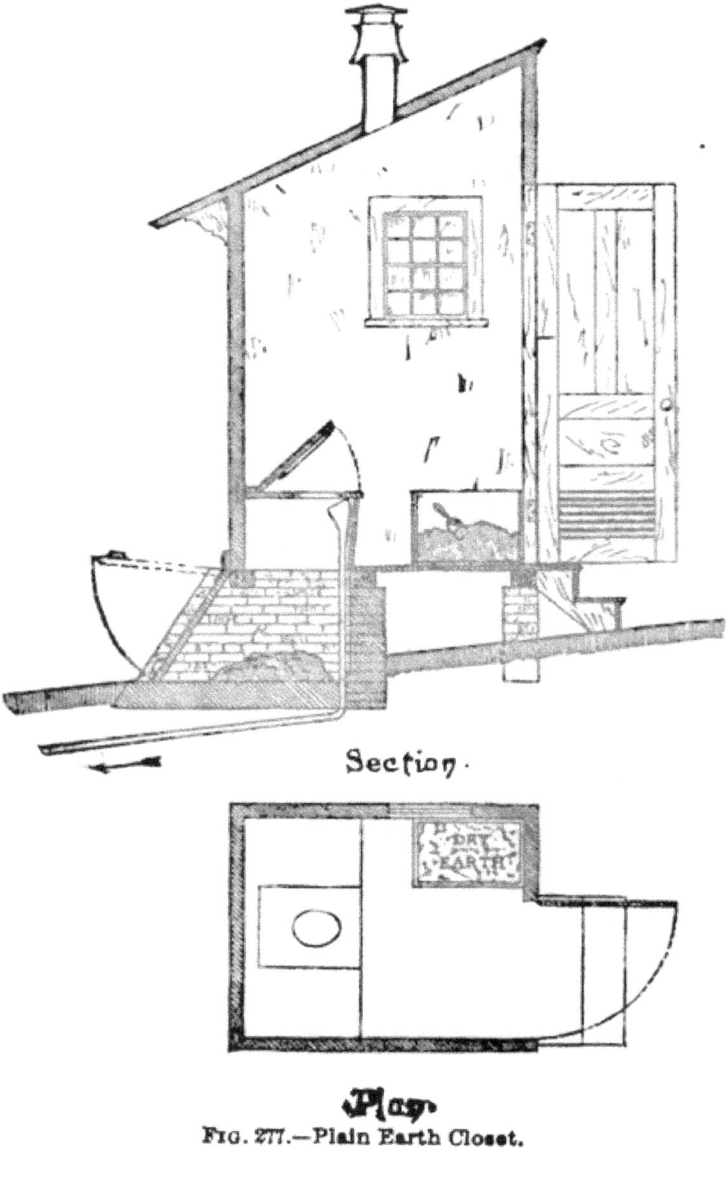

FIG. 277.—Plain Earth Closet.

the cemented vault, for this would be sure to create a nuisance.

We have shown under and in front of the seat in Fig. 277, a funnel, intended to catch and remove the urine by means of a small pipe leading to the slop-water tank (shown in Fig. 276). Although we are aware that it is impossible to separate *all* urine from the excreta, we are strongly inclined to believe that such a separation will tend to lessen the possibility of an earth-closet becoming offensive.

We must now resume the description of various constructions of the sewage tank. For larger buildings, a *tight* cesspool of dimensions sufficient to hold one or two days' sewage, must be built. Its outlet may be closed by a gate, operated by hand labor. As this may not always be done with regularity, an *automatic* arrangement for the discharge is preferable.

The capacity of the tank should be larger than the capacity of all absorption tiles. Its whole contents should be suddenly delivered into the pipes, whereby *all the rows of tiles are uniformly charged.* Thus, the *whole* of the absorption field is brought into use each time the tank is emptied. The purification begins immediately, the clarified liquid soaks away into the ground, the impurities being retained by the earth filter, where they are destroyed by oxidation, air enters the pores of the soil and prepares it for future use, while the tank is gradually filling for the next discharge.

An important caution for all cases where the contents of water-closets are to be disposed of combined with slop water, is to *intercept all solids and fatty waste matters*, which should not be discharged with the liquid sewage into the absorption drains, as they would, in a short time, clog these, and also interfere with the action of the flush tank. An intercepting chamber must be built between the house and the flush tank, such as shown in Fig. 281.

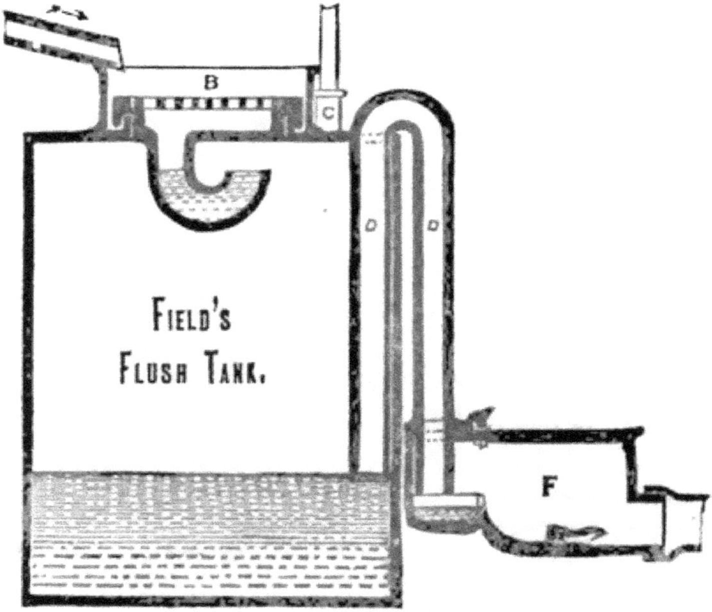

FIG. 278.—Field's Flush-tank.

This will, in a certain sense and to a certain degree, be a cesspool; its contents, however, are frequently changed, it can be kept of small dimensions, and its emptying and cleaning (a matter which must by no means be neglected) is much more easily effected. It should be built of best hard burnt brick, set in pure Portland cement, and the tank rendered *perfectly tight.*

The automatic discharge of the sewage tank can be effected either by means of a siphon or else by a tumbler tank.

Rogers Field's small siphon tank is shown in Fig. 278. It is made both in earthenware and in cast-iron, and holds about 40 gallons. This tank is intended for the disposal of slop-water, but may also be used for flushing house drains.

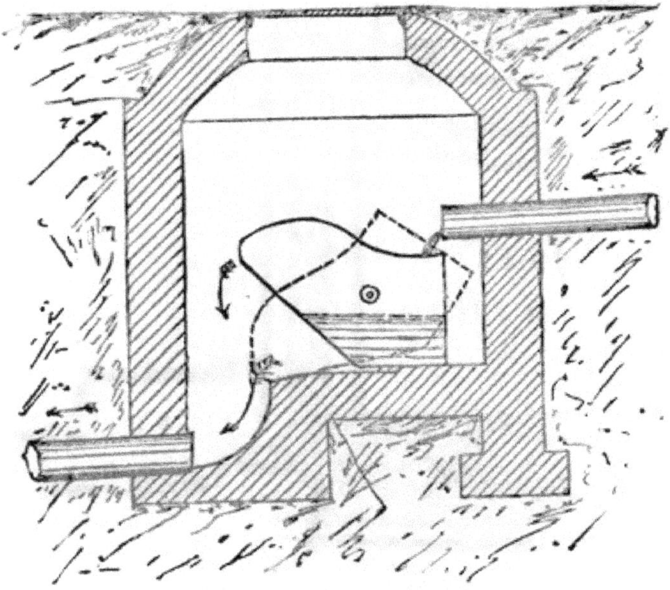

FIG. 279.—Tumbler Tank.

A tumbler tank for slop-water is shown in Fig. 279, and a combination of a siphon and a tumbler box (Isaac Shone's house sewage ejector) in Fig. 280.

Any of these tanks may be used for houses having water closets, if an intercepting chamber is placed between the house and the tank.

A larger tank, built of brick, is shown in Fig. 269, and on a large scale in Fig. 281. A represents the house drain, B the intercepting chamber, C the flush tank, DD are tight iron covers, E is a deposit of sludge in the intercepting chamber, F is the overflow pipe from it to the flush tank which dips at least 12 inches into the liquid, to prevent any solid matter or greasy scum from being carried over into the flush tank. H is the annular siphon which

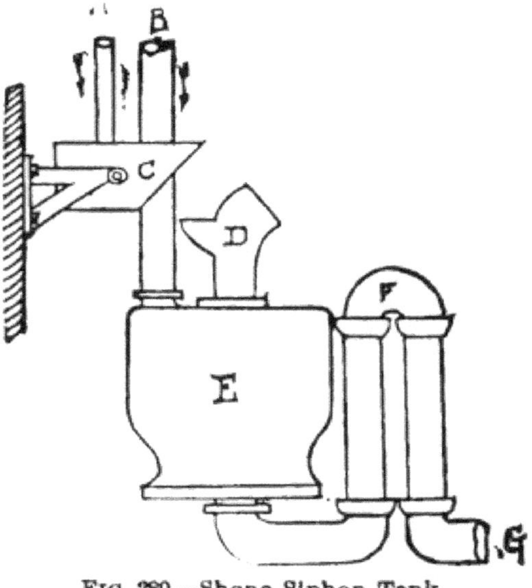

FIG. 280.—Shone Siphon Tank.

discharges the contents of the tank at regular intervals automatically. Its details are shown in Fig. 282. K is a screen of iron wire, M is the weir which starts the siphon, L is an inspection pipe over the wier, closed at the surface by a trap screw P, N is the drain leading to the irrigation field.

The operation of this tank may be described briefly as follows:

As soon as the tank is filled up to the level XX, the water begins to overflow through the inner limb of the siphon. With the sudden discharge of a bath or wash-tub, enough water generally overflows to seal the siphon at its bottom, as the water cannot pass out through the weir M as quickly as it rushes down the siphon. The descending column of water carries air with it and thus establishes a partial vacuum in the siphon, whereupon the air pressure in the tank forces enough water into the siphon entirely to fill it. Thus the siphon is started and continues to discharge the contents

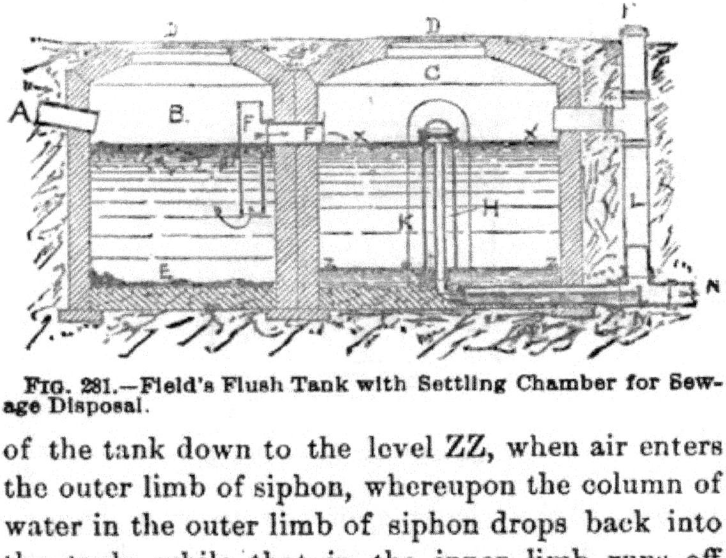

FIG. 281.—Field's Flush Tank with Settling Chamber for Sewage Disposal.

of the tank down to the level ZZ, when air enters the outer limb of siphon, whereupon the column of water in the outer limb of siphon drops back into the tank, while that in the inner limb runs off through the weir. Air enters here and completely breaks the siphon, while the tank is gradually filling up.

To protect the siphon from obstructions through paper or grease carried over from the catchment apparatus, it is advisable to place around it a net of galvanized iron wire of about ½ inch mesh. Even with this protection the siphon needs frequent cleaning off by means of a hose, otherwise serious stoppages, especially in the notch of the weir, will occur. To remove these obstructions, an inspection or lamphole is placed, as shown, directly over the weir.

It should be mentioned that the annular siphon, Rogers Field's invention, is patented in England, and its application to tanks for flushing sewers as well as to sub-surface irrigation, is controlled in

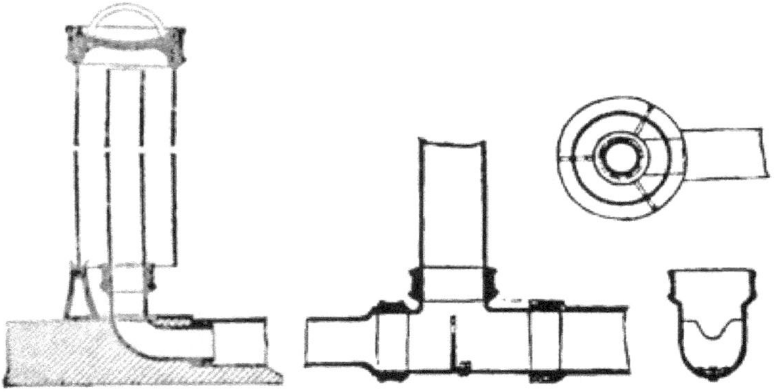

FIG. 282.—Field's Annular Siphon.

this country by the Drainage Construction Company, of which Col. Geo. E. Waring, Jr., is consulting engineer.

Fig. 281 shows only one method of construction of a flush tank with Field's siphon for sewage disposal. It may easily be modified and possibly improved.

The disposal of household wastes is a subject which might well demand a treatment in a special volume, and since it was not our intention to describe with great minuteness all details of the systems of sewage disposal for country houses, we have omitted to speak of the proportion between size of house and capacity of tank, between the latter and the size of the irrigation field, between the size of tank and number of feet of distributing drain tiles, between the character of soil and the distance between the rows of tiles, etc., all of which are details requiring judgment, skill and experience on the part of the designer of such a system. Local conditions will largely determine the design and arrangement of the tank and the laying out of the irrigation field.

Suffice it to say that there exists in no case a sound excuse for storing the human filth in the usual leaching, unventilated cesspool placed in close proximity to the household, the best means for breeding or multiplying disease germs and spreading disease in case the seed should reach it. A mass of putrescent human filth stored beneath or near a dwelling has well been compared to a powder magazine, for one single little spark—a germ in the stool of a typhoid fever patient—may suffice to create vast harm and destruction.

To contribute his share in the prevention of "*preventable*" disease has been the author's aim in writing these hints. His hope is that in a near future we may find in and around every human habitation, in the city and in the country, "*pure air, pure water, and a pure soil.*"

THE END.

www.ingramcontent.com/pod-product-compliance
Lightning Source LLC
Chambersburg PA
CBHW021504210326
41599CB00012B/1126
9783744678360